*The World's
Most Famous Book
on Glassmaking*

The Art of Glass

by
Antonio Neri

translated into English

by
Christopher Merrett

Edited by
Michael Cable

The Society of Glass Technology
2006

Art of Glass by Christopher Merrett edited by Michael Cable
This is the first volume in a chronological series showing in the words of authorities of various epochs and several countries how understanding of glass-making developed over four centuries. The series so far published is:
1. The art of glass by Christopher Merrett (1662)
2. Bosc D'Antic on glass making (1758–80)
3. Early 19th century glass technology in Austria and Germany (1820–37)
4. Apsley Pellatt on glass making (1820–49)
5. Georges Bontemps on glass making (1868)

© Michael Cable 2006

Published by the Society of Glass Technology

The objects of the Society are to encourage and advance the study of the history, art, science, design, manufacture, after treatment, distribution and use of glass of any and every kind. These aims are furthered by meetings, publications, the maintenance of a library and the promotion of association with other interested persons and organisations.

Society of Glass Technology
Unit 9, Twelve O'clock Court
21 Attercliffe Road
Sheffield S4 7WW
UK

Registered Charity No. 237438

Web site: http://www.sgt.org

ISBN 0 900682 37 X

CONTENTS

Preface	1
A notable British seventeenth century contribution to the literature of glassmaking. *W.E.S. Turner*	3
The Art of Glass	39
Dedication to Lord Antonio Medici by Neri	43
To the Curious Reader	45
Dedication to Robert Boyle by Merrett	53
To the Ingenuous Reader	55
To Avoid Our Author's Repititions	58
The first Book: Chapters 1–XXXVI	59
The second Book: Chapters XXXVII–XLIV	117
The third Book: Chapters XLV–LX	145
The fourth Book: Chapters LXI–LXXIV	163
The fifth Book: Chapters LXXV–XCII	181
The sixth Book: Chapters XCIII–CVII	203
The seventh Book: Chapters CVIII–CXXXIII	217
Index	259
Merrett's Observations on Neri	
The Epistle to the Reader	263
Of the Furnaces	297
On the first Book	308
On the Author	367
An Account of the Glass Drops	411
An Appendix	421
Corrections	424
General index	427

PREFACE

There can be little doubt that Neri's book deserves the accolade accorded to it by the title of this volume: it was translated into five other languages and printed no fewer than twenty times over the course of three centuries but has been neglected for a century or more. Why it still deserves study is made clear in the essay by W.E.S. Turner (his last publication) which introduces it here.

W.E.S. Turner OBE DSc FRS (1881–1963) was in 1904 appointed assistant lecturer in Physical Chemistry at the University College of Sheffield where he quickly showed outstanding talents as an organizer and an interest, unusual among academics in those days, in how the University could help local industries. He was soon giving very popular lectures on Chemistry for Metallurgists.

When the first World War began he suggested that the University set up a committee to advise and assist local industries and he became its secretary. It quickly became clear to him that the glass makers particularly needed advice about the application of science to their operations and, moreover, were willing to seek it. As a consequence in 1915 he wrote a report on the Yorkshire glass industry that recommended the setting up of a *Department of Glass Manufacture* at the University. That report was quickly accepted and within six months Turner, not surprisingly, was asked to set up and direct the new Department which was soon renamed the *Department of Glass Technology*, a term that he probably invented. It was not then his intention to abandon chemistry but he quickly became totally committed to all aspects of glass and devoted his considerable energies to the world of glass throughout the rest of his long career.

Within a year he had also established the Society of Glass Technology (with the particular help of Frank Wood of Wood Brothers, Barnsley) and, in the 1930s, was one of the founders of the International Commission on Glass. His abilities and achievements were recognized by many honours in various countries and not confined to the world of glass.

He retired from the University in 1945 but was still a frequent visitor to his old Department when I became a student of Glass Technology in 1952; from 1954 we would often meet in the Department Library and occasionally converse. He was probably engaged in the historical researches that were then his main interest, one of the fruits of which is the magisterial essay reproduced here. His encyclopaedic knowledge of the literature was often of value to me. On several occasions I was asked (not directly by Turner himself but by Michael Parkin, one of Turner's loyal lieutenants) to melt glasses of rather odd compositions usually with the instruction "do not exceed 1200°C"; only later did I realize for whom the results were intended. When I travelled to the USA in 1959 to take up a research appointment at Massachusetts Institute of Technology, he gave me letters of introduction, which proved very valuable, to several of his friends there. This book is a small act of homage to him.

Although not a facsimile of Merrett's volume, the main text follows that in content and style as closely as the available facilities allowed: the original printer's ornaments have been reproduced and Merrett's pagination retained. The only differences are that a few floridly decorated capital letters are omitted and some errors corrected. Besides the *errata* appended by Merrett himself, other corrections considered to be obvious have been made. Just two of these made here for the first time deserve comment: (1) in his essay Turner noted an obvious and surprising mistranslation by Merrett (*conterie* as *counting houses*) and (2) Merrett's list of the characteristic properties of glass describes the perfect elasticity of glass fibres by the term *motum rectitutionis* which the Latin translation of 1668 changed to *motum rectitudinis* but the more intelligible correction of a simple printer's error surely is *restitutionis*. These corrections are all underlined and also listed at the end, so any reader wishing to do so can easily reconstruct exactly what Merrett's original shows.

The University of Sheffield　　　　　　　　　　　　　　　　*Michael Cable*

A NOTABLE BRITISH SEVENTEENTH-CENTURY CONTRIBUTION TO THE LITERATURE OF GLASSMAKING

W.E.S. Turner

Reprinted from *Glass Technology,* **3** (6), 201–13 (1962).

In the entrance hall at 'Elmfield' there is a large plaque intended to fulfil two purposes: first to place permanently on record the names of those generous donors who made possible the erection of the fine buildings[†] there and secondly to remind all who come and go through this entrance hall of some of the outstanding contributions to the art and technology of glass made by the British glass industry during the seventeenth and eighteenth centuries.

These achievements include the adoption of coal for the melting, fritting and annealing of glass; the construction of efficient furnaces whereby smoke could be kept at a low level and temperatures at a relatively high level; the closed melting pot; and a lead-containing glass suitable for making crystal vessels. This last-named development is usually ascribed to George Ravenscroft and the date of its realization as 1675.

It has not hitherto been sufficiently recognized that developments in glass making techniques were greatly stimulated by the publication in 1662 of a book entitled *The Art of Glass*. This owes its existence

---

[†] *Editor's note.* The Elmfield laboratories were specially built in 1937–39 for the *Department of Glass Technology*, (later the *Department of Ceramics, Glasses and Polymers*) of the University of Sheffield and its home until 1993; Turner raised most of the funds for the original buildings him-

to the labours of two men, Antonio Neri and Christopher Merrett,* who not only translated Neri's work into English but added extensive 'Observations' and notes of his own occupying at least one half of the book. It is generally agreed that Neri's Italian work of 1612 was the first account of glassmaking which accorded with modern understanding and presentation, while Merrett's was the first account of the subject in the English language.

While I have not ascertained in what month of 1662 *The Art of Glass* was published†, it seemed to me very fitting that this Society should, at its Annual Meeting in 1962, take the opportunity of celebrating the three hundred and fiftieth anniversary of Neri's Italian work and the three hun-

---

* There was considerable looseness in the sixteenth and seventeenth centuries in the spelling of names. Thus Merrett at baptism was registered as Merret, the son of Christopher Merret, churchwarden of St Peter's Church, Winchcombe. The father was also referred to in another record by the spelling Merrit. The Royal Society of Physicians invariably spelt the name 'Merrett' whereas the Royal Society and most foreign authors wrote 'Merret'. Thomas Birch in his *History of the Royal Society* referred to him more than one hundred times in his summaries of all the meetings held from 1660 to 1674 and invariably employed the spelling 'Merret'. I have adopted the spelling with two ts advisedly because because that was the form with which Merrett signed documents associated with his post of Librarian to the Royal College of Physicians and as author of papers. What seems most odd is that the Royal Society should ignore the spelling with the double 't' which Merrett himself used when appending his name to the list of signatures of those who accepted membership at the first meeting of the Society held on 6 December 1660. These signatures are preserved and attached to the statement of the *Charter Book* of 1663 and the double 't', despite the small writing, is clearly decipherable.

† *Editor's note.* Probably September or October.

THE
Art of Glaſs,
WHEREIN
Are ſhown the wayes to make and colour Glaſs, Paſtes, Enamels, Lakes, and other Curioſities.

Written in *Italian* by *Antonio Neri*, and Tranſlated into *Engliſh*, with ſome Obſervations on the Author.

Whereunto is added an account of the Glaſs Drops, made by the Royal Society, meeting at *Greſham College*.

LONDON,
Printed by *A. W.* for *Octavian Pulleyn*, at the Sign of the *Roſe* in S*r*. *Pauls* Church-yard. *MDCLXII.*

Figure 1: Merrett's original title page

dredth anniversary of Merrett's translation and substantial extension.

It is interesting that although the title page of this book does mention Antonio Neri, the name of Christopher Merrett does not appear either here or anywhere else in the whole work. Despite his modesty there never was doubt about the author's identity, for the work had been undertaken by Merrett as one of his contributions to the varied activities of the group of scientific men to which he belonged, who from 1645 onwards met regularly but informally in London, and ultimately in 1660 founded the Royal Society. The nearest indication to the name of the translator and author of the 'Observations' is the 'C.M.' which he appended both to his general preface and to his dedication of the book to Robert Boyle where he wrote:

> ... you were the principal cause that this Book is made publick, by proposing and urging my undertaking of it, till it came to a command from the most Noble Society, and serious indagators of Nature, meeting at Gresham College, whose desire I neither could nor ought to decline.

After publication of the book in 1662, the Royal Society at its meeting on 8 October of that year recorded the following Minute:

> It was ordered that the thanks of the Society be given to Dr. Merret for his pains in translating the Italian discourse *L'Arte Vetraria* upon the motion and desire of the Society.

A further Minute at the meeting on 22 October records that Dr. Merrett was desired to bring in the *Materia Vitraria* mentioned in his book. This he did at the meeting on 5 November after which he was asked to leave a part of each of the materials with the Society.

Having now introduced my characters I propose to give an outline account of the life and of the qualifications of each for the parts which they played, to comment on selected portions of each text, and finally to discuss the influence which their joint labours had in stimulating the interest of glassmakers and scientists not only in England but throughout western Europe.

Antonio Neri, d. 1614

Brief biography

There is a short account of the life and work of Antonio Neri in the *Biographie Universelle*, volume 31, Paris, 1822, without any clue as to sources of information and no date of birth or death. It could be an imaginative write-up based on the few references to his travels and work in different centres given by himself. He was generally referred to as a Florentine monk and his pious expression in concluding the preface and other passages in the text consort well with such a profession. He was but one of several monks who during several centuries were important contributors to a knowledge of glass.

The preface to his book *L'Arte Vetraria* was written in Florence and signed on 6 January 1611. Publication took place during the following year in Florence, the book being dedicated "To the most Illustrious and Excellent Lord Don Antonio Medici". Merrett, at the outset of his "Observations on the Epistle to the Reader", said that in all his reading he had found references to Neri in the works of only two other authors. Certainly Merrett does not appear to have been aware that Neri died in 1614, the year of his own birth. The record of the year of Neri's death appears in an entry in J. C. Poggendorf's *Biographische literarisches Handwörterbuch*, Leipzig, 1863, and it is doubtless from this source that Duncan recorded the date of death in his *Bibliography of Glass*.

From Neri's book we are able to glean a little information about the centres in which he had worked. Four such centres are in fact referred to in the text: Pisa*, Florence[†], Antwerp[#] and Flanders[$]. It would seem, from these references, that whatever his attachment to Florence and to his patron there, the place where a substantial part of his work was carried out was Pisa. Here, according to his claims, he not only made several different glasses "many times", but also designed a furnace for the more efficient calcination of copper and invented two ways of making a lake from cochineal, one for painters and another for colouring glass.

When referring to his activities in Antwerp he described a seagreen glass he had "often" made there, so that he must have spent some con-

---

*   pp. 13, 41, 46, 49, 52, 53, 149, 177, and 179.
[†] pp. 40 and 74.
[#] pp. 49–50, 84, 120, and 171.
[$] p. 78.

siderable time in that city. Other activities in Antwerp included the preparation of ultramarine and the working out of one of his processes for making a glass resembling chalcedony. This particular glass seems to have brought him much satisfaction, and friendship with several gentlemen glassmakers and nobles. Thus at Florence he instructed "the brave Gentleman" Nicolao Landiamo in the working of this glass at the furnace in Casino, and at Antwerp in January 1609 he instructed Signor Philippo Ghiridolpho in working the chalcedony glass made by his third method. In the house at the same time was a Portuguese gentleman of wide learning, Signor Emanuel Nimenes a "Knight of the Noble Religion of Saint Steven… and Citizen of Antwerp … ". Two vessels of this chalcedony glass were presented to and pleased the Prince of Orange.

L'Arte Vetraria

Turning now to the contents of Neri's book it would be quite impracticable to refer in a single paper to its many features, and I propose to confine myself to an exposition of the conditions under which he was working and writing, and then after a brief survey of the text to select a few sections for special comment.

Circumstances relating to publication of the book. As the sixteenth century moved forward chemistry or alchemy moved steadily away from the views of Aristotle that the constitution of matter was dependent on the four elements fire, earth, air, and water transformable one into the other. To an ever-increasing extent chemistry became concerned with the production of new substances useful in medicine and in the industrial and decorative arts. This trend away from theory towards utilitarian purposes is to be seen in the system of alchemy or chemistry of Andreas Libavius, a distinguished German chemist (1540–1616). His book *Alchemia* (two volumes; 1st ed. 1597, 2nd ed. 1606) was described by J. C. Poggendorf as the first systematic work on chemistry. In it the end-product was classified according to the process by which it was obtained: solution, precipitation, crystallization, fusion, distillation, etc.

How then did the glassmaker proceed at that date and time? The answer is that he followed a traditional procedure* which can be traced back to the earliest times B.C., down through the period when Neri was working at the beginning of the seventeenth century, and indeed into the nineteenth. Two ingredients were regarded as essential to make glass. The first was rock crystal, silica rock, or sand reduced to powder and sieved to the fineness of flour and, in that condition called "tarso" by Neri, a term for which Merrett found no English equivalent and therefore retained, and the second, alkaline glass salts usually derived from the ash of burnt plants or trees. These ingredients were weighed, mixed, and heated in two stages. In the first, the mixed substances were raked together for several hours on the hearth of a reverberatory furnace called a *calcar* at a temperature too low to produce complete fusion and liquefaction, but sufficiently high to lead to granulation and at least a semi-vitreous condition. The product of this treatment, known as *fritt*, was usually kept as stock material to be charged as required into melting-pots in fixed amounts. In the second stage, prescribed quantities of other ingredients were added to the fritt for producing special characteristics, or for colouring or decolorizing, and the whole heated and treated until homogeneous glass was obtained.

Despite this apparent simplicity of principle the glassmaker had before him a complicated task. Even when by using powdered rock crystal or silica rock he avoided the complications due to impurities in sand, the variation in the composition of the alkaline glass salts affected their alkalinity, and consequently the rate and degree of reaction with the

---

* For a fuller exposition of this procedure see the author's "Studies in ancient glasses and glassmaking processes", Part V. Raw materials and melting processes. *J. Soc. Glass Tech.* (1956), **40**, 277 T.

tarso. The proportion of the two ingredients was therefore dependent on the purity of each and on the judgement of the glassmaker. One method of evaluating the alkalinity of the ash, referred to both by Neri and by Merrett, was to taste it. Both of them, however, commend as the safest method a series of fusion experiments in small crucibles in which the amounts of silica and ash could be varied to ascertain the proportions which would readily produce a satisfactory glass. Neri (pp. 7–8), using ash specially purified as described in his opening chapter, found favourable results when he employed 200 lb. of prepared tarso and 130 lb. of purified ash. The result first of fritting and then of melting a product of this basic composition yielded his so-called crystal glass.

If, however, common glass or green glass was the aim of the glassmaker, the ash did not need to be purified rigorously; indeed, for green glass there would be no purification. It followed accordingly that the proportion of this cruder type of ash required would be very much higher and that the ash would carry with it into the glass all the impurities corresponding to its crude state.

In his preface "To the Curious Reader" Neri at the outset dwells on the wonderful nature of glass and its service to man in so many diverse ways, including science and medicine. One might have expected reference to the making and treatment of glass suitable for these purposes, but his book is confined to those types of glass which Venice and other Italian centres had made peculiarly their own; crystal drinking-glasses, objects for lighting and display, artificial jewellery, enamels, and painted glassware. These types of glass had brought Venice high reputation and wealth, and other countries of Western Europe were eager for information about them.

Neri had clearly come to realize that the finest of glasses, crystal-clear, coloured, or opaque could only be regularly produced from constituents which had been prepared and purified to the highest standards attainable. To this end all the knowledge and the concentration of a trained chemist were needed and Neri appears to have had the qualifications essential for the task. His instructions were clear and meticulous, and those of basic or fundamental character were repeated with every separate glass-melting operation. Merrett, having translated the whole of Neri's original text, said that the tedious repetition of instructions caused him such nausea that after seeking literary advice he collected them under five headings. labelled "To avoid our Author's Repetitions, Observe", and inserted them immediately before Chapter 1. These instructions insisted on the use of hard dry wood to heat the furnaces and to avoid smoke; the thorough grinding together of any mixture of colouring agents before addition to the glass; the stage by stage addition of various metals, etc.

A survey of the text. Neri's text is divided into seven Books, presumably written at different periods though no doubt revised from time to time. The chapters, 133 in all, are numbered consecutively throughout the work and each has a special theme. In general the text is devoted to (1) the preparation of the foundation materials; tarso, the alkaline salts derived from plant ash, and tartar from red wine (potassium hydrogen tartrate); (2) other foundation materials such as lead oxide and tin oxide; (3) colouring oxides including those of iron, copper, manganese, and cobalt (zaffre). The methods of extracting colours from flowers and plants to produce lakes for painting is also described. Several chapters are occupied by instructions for the preparation of the acids used for dissolving metals, and a much larger number of chapters is taken up with directions for melting numerous kinds of glasses, enamels, and artificial gems.

References to furnaces and processes are few and casual, a knowledge of these aspects of the subject being taken for granted; but there are a number of passages in the text which reveal Neri's practical knowledge of glassmaking. Thus, in addition to the instructions which Merrett found so tedious, the preparation of fritt in the calcar provides an example. The instruction here (p. 8) is that the fritting furnace shall be made moderately hot before introducing the mixture of tarso and ash, by which time smoke should have ceased and the temperature should be high enough to make fritt. Neri observes that if the tarso and ash

Figure 2: A furnace, originally from Agricola, *as reproduced in the Latin translation of Neri (1668).*

be put into the Calcar when it is cold, Fritt will never be made of them.

The reason is that the high thermal insulating power of the finely divided mixture would effectively prevent the hearth of the calcar from becoming hot enough to make granulated fritt. Nearly forty years ago the Department of Glass Technology found that if flames were played on a batch pile in a tank furnace the centre remained powdery fifteen to twenty minutes after the batch had been introduced.

I propose, because of lack of space, to confine detailed comments on the text to Books I and IV.

Book I is the longest of the seven, comprising thirty-six chapters and occupying pages 1–58, or more than one-fourth of the whole work. Six chapters are devoted to the preparation and purification of the alkaline salts, one to the preparation and evaluation of tarso, four to the making of crystal fritt, thirteen to methods of producing colouring oxides by the calcination of metals, one to general observations on the making of coloured glasses, and eleven to making coloured glasses from the reagents mentioned.

The fundamentally important item is the purification of the alkaline salt, starting with Polverine or Rochetta by which names the raw ash from the Syrian and Egyptian coasts was known, or Barillia, the Spanish material from Alicante. The crude ash was crushed small and boiled repeatedly with successive quantities of water in brass cauldrons, the extracts being ladled out into earthenware bowls to settle, and the clear liquid evaporated until crystallization occurred. Before boiling, a quantity of partly calcined tartar was placed in each cauldron to act as a reducing agent; this was claimed by Neri to have been his secret process to produce "… more, and whiter salt …". The salt as it crystallized was removed by a scummer (Figure 5, B), drained and dried, then heated in the calcar until hard. It was finally crushed and sieved to produce particles the size of grains of wheat.

Neri states that from three hundred pounds of the raw Polverine he was accustomed to get between eighty and ninety pounds of the finest ash. Whilst the advantage of the purification process was that it produced glass of more brilliant appearance, the fritt and the glass had much lower stability. In regard to the stability of the fritt, Neri (Chapter 11) says:

> This Crystall Fritt must be kept in a dry place, where no moisture is, for from moist places, the Fritt suffers much, the salt will grow moist, and run to water, and the Tarso will remain alone, which of it self will never vitrifie: neither is this Fritt to be wetted, as others are.

In his "Observations" on this particular point Merrett (p. 211) comments:

> "furthermore in the finest Glasses, wherein the salt is most purified, and in a greater proportion of salt to the sand, you shall find that such Glasses standing long in subterraneous and moist places will fall to pieces, the union of the salt and sand decaying".

He also hints that the legend about the finest glasses breaking when poison was dropped into them might possibly be traced back to their low stability to anything slightly corrosive, whether atmospheric moisture, mineral acid, or poison. Anyone who has examined the fine specimens of Venetian glassware of the sixteenth and seventeenth centuries will recall the corrosion films which are often present in varying thicknesses. Neither Neri nor Merrett had more than a glimmering of the fact that the big loss in weight on purification of the ash was due largely to the removal of stabilizing constituents, principally lime and magnesia with some alumina, titania, and silica. Moreover, although potash was a term which was in general use, a clear differentiation between sodium and potassium salts was not worked out until 1762, a hundred years after the publication of Neri–Merrett.

Turning now to Book IV, this has fifteen chapters wholly given over to lead glasses. Far from being a comparatively modern glass constituent, lead was used in glazes in very early times in Babylon (1700 B.C.) and in glasses at Nimrud (800–600 B.C.). Moreover, because the starting point was usually metallic lead, which was first converted to the oxide by calcination, lead can be said to be the first identifiable element used in glassmaking. It continued to be used throughout the centuries, particularly for mosaics and gems. A drawback to its use was its corrosive action on the melting pots.

Neri's first mention of lead as a constituent occurs as early as Book I, Chapter XXXV, whilst in Book III, Chapter LII, a lead batch coloured black by calcined steel and hammer scale is given, and in Chapter LIV, a process for making a milk-white glass called *Lattimo* is described, based on the addition of a mixture of calcined lead and tin. After Book IV most of the recipes are based on lead oxide as the major constituent. In Book V, utilizing the experience of Isaac Hollandus, the preparation of artificial gem stones is described; in Book VI, enamels for goldsmiths; and some miscellaneous glasses are included in Book VII.

Book IV is historically important because of the work of George Ravenscroft. From its fifteen chapters I select three to tell their own story of the substantial knowledge of lead glasses which Neri claimed to possess.

Chapter LXI

The glass of Lead known to few in this Art, as to colours, is the fairest and noblest glass of all others at this day made in the furnace. For in this glass the colours imitate the true Oriental gems, which cannot be done in Crystal, nor any other glass. 'Tis very true, that unless very great diligence be used, all sorts of pots will be broken, and the metall will run into the coals of the furnace. Observe my rules in all these glasses made of Lead exactly, and you shall avoid all danger. This business principally consists in knowing well how to calcine Lead, and to recalcine it also a second time; For by how much 'tis better and more calcined, by so much the less it returns to Lead; Again, and by consequence the less breaks out the bottom of the pot. Secondly, cast the metall into water, and separate carefully the Lead from the glass, even the least grains of it. This glass of lead must be cast into the water by little and little, to make a better separation, for the least Lead remaining breaks out the bottom of the pots, and lets all the metall run into the fire.

These two rules our Author repeats almost in every Chapter of this Book, and these following also,

The pots and Lead must not have too much heat in the furnace, neither must the metall be wrought too hot, and the Marble whereon 'tis wrought must be of the hardest stone, and must be wetted else the marble will break and scale.

Chapter LXII
To calcine Lead

At first Calcine Lead in a Kil as potters do and in great quantity. Usually in two days they calcine many a hundred pound of Lead. In calcining observe that the Kil be not too hot, but sufficiently heated onely to keep the Lead in fusion, for otherwise twill not be calcin'd. When the Lead is melted it yields at the top a Yellowish matter. Then begin to draw forwards the calcined part with an Iron fit for the purpose, always spreading it in the internal extremity of the Kils bottom, which should be of soft-stone, which will bear the fire. And the Kil must have a declivitie towards the mouth, which I pass by as a thing well known. When 'tis calcined once it must be put, and spread a second time in the Kil, to reverberate in a convenient heat, always stirring it with an Iron, and that for many hours, till it come this second calcination to a good Yellow and be calcined. Then serce all in a fine serce, and what passeth not the serce recalcine it with new Lead. This is the way to calcine Lead in great quantity to make thereof store of Potters ware.

Chapter LXV
Glass of Lead of a wonderful Emerald colour

Take of *Polverine Fritt* 20 pound, Lead calcined 16 pound, serce these two powders first by themselves, then, when well mixed, put them in a pot not too hot, and at the end of 8 or 10 hours they will be melted, then cast them into water, and separate the Lead. Put them a second time into the pot, and in 6 to 8 hours they will be melted, then cast them into water and separate the lead. This being twice done the metall will be freed from all the Lead, and all the unctuosity which calcined Lead and Polverine give it, and will acquire a most bright and shining colour, and in few hours 'twill run and become very clear, then give it brass thrice calcined (made as in Chap. XXVIII), six ounces, and therewith mix a peny weight of *Crocus Martis* made with Viniger, put in this mixture at six times, always mixing well the glass, and taking at each time the intervall of saying the Creed. Let this glass settle an hour, then mix and take a proof thereof. When you like the colour let them incorporate 8 hours, then work them into drinking glasses, which will appear in a colour proper to the Emerald of the old Oriental rock, with natural shining and glittering.

Let this glass stand in a pot when sufficiently coloured, till It hath consumed all the dregs, and till it be perfectly refined, and then 'twill be so like the natural Emerald that you can hardly know one from the other.

It could hardly be gainsaid that here was valuable information which should have been of great usefulness in the development work of Ravenscroft and his co-workers. The reader interested in this aspect of the subject should read Chapter V. entitled "The Rise of English Crystal", particularly pp. 140–58, in W. A. Thorpe's *English Glass*, where the relationships between the London Glass Sellers and Ravenscroft are discussed and the successive stages mentioned by which complete technical success was ultimately achieved. Thorpe suggests that the incorporation of lead into Ravenscroft's glass may have arisen by noting its effect in lead glazes applied to English earthenware. It is hardly likely that Ravenscroft, engaged as he was on a most important commercial venture, would disclose his technical procedures. But as he did not begin his experiments until after 1670 it would be strange indeed if he, either before making them or during their course, did not become acquainted directly or through his English or Italian friends with the Neri–Merrett *Art of Glass* of 1662 with its Royal Society approval, the new Italian editions of Neri in 1661 and 1663, or the Latin editions of Neri–Merrett published in Amsterdam in 1668 and 1669.

Christopher Merrett (1614–1695)

While Neri during his lifetime had pursued his quiet way as a student and teacher of the art of glassmaking, Christopher Merrett in his day was a much more active figure in medical and scientific circles, being a distinguished physician, a leading botanist, and a Founder Fellow of the Royal Society in 1660. He might have attained still greater fame but for misfortunes which had their origin in the Plague of London and in the severe losses sustained by the Royal College of Physicians and its library as the result of the Great Fire of London while he held the Librarianship.

There are excellent biographical notices of him in Wood's *Athenae Oxonienses*, IV, 430–432; in W. Munk's *Roll of the Royal College of Physicians of London*, 2nd edition, 1878, Volume 1; by G. S. Boulger in the *Dictionary of National Biography*, 1894, 37, 288; and a recent sympathetic study of his career by Professor Sir Charles Dodds (*Proc. R. Soc. Med.* (1954), 47, 1053), Christopher Merrett. F.R.C.P. 1614–1695, First Harveian Librarian.

Christopher Merrett was born at Winchcombe, Gloucestershire, on 16 February 1614, receiving the same Christian name as his father. In 1631 he was admitted a student of Gloucester Hall, Oxford, an old foundation which at a later date became Worcester College. He removed from Gloucester Hall to Oriel College about 1633, graduated B.A. in January 1635, and then returned to Gloucester Hall to devote himself to medicine, acquiring the degree of M.B. in June 1636, and after gaining considerable experience achieved the degree of M.D. in January 1643. Having settled in London he became a Fellow of the Royal College of Physicians in 1651, and there opened out to him a career of great promise. He was Goulstonian Lecturer in 1654, a Censor of the college seven times between 1657 and 1670, an Elect of the college, and its first Librarian.

His nomination to the Librarianship arose from his friendship with William Harvey who, as the discoverer of the circulation of the blood, had acquired a European reputation. In 1614 the Royal College was removed to new premises at Amen Corner where there was adequate space to house a library. Merrett appears to have been a successful librarian, the library greatly increasing in size under his guidance. A complete catalogue of the library contents before the Great Fire was prepared by him and printed in 1660. He also wrote a number of works associated with the practice and dignity of the medical profession in some of which he sharply took to task some whose methods smacked of quackery; his candour could likely have brought him enemies.

We are concerned in this account with Christopher Merrett's contribution to *The Art of Glass* and this came about as a result of his association with the founders of the Royal Society. I propose, therefore, to speak of this association and of the very active part which he played for a period of at least twenty-five years in that distinguished body.

In his *Defense of the Royal Society*, published in 1678, Dr. John Wallis, a mathematician of great repute, recorded that 'about the year 1645 a number of worthy persons residing in London who were inquisitive into natural, and the new experimental, philosophy, agreed to meet weekly on a certain day to discourse upon such subjects'. They included Dr. John Wilkins, Dr. John Wallis, several professors of Gresham College distinguished in mathematics and astronomy, and a number of physicians associated with the Royal College of Physicians, among them Dr. Jonathan Goddard, Dr. George Ent, and Dr. Christopher Merrett. They met, according to circumstances, at Dr. Goddard's lodging, at the 'Bull's Head', Cheapside, or at Gresham College. No record appears to have been kept of the subjects discussed or of conclusions reached but it cannot be supposed that they underwent any abrupt change in character on the establishment of the Society in 1660. From that time onwards records of the proceedings of meetings were systematically kept and from them was ultimately compiled – a century later – the *History of the Royal Society*, by the Secretary, Dr. Thomas Birch. On its Incorporation in July 1662 the Society was given the title of 'The Royal Society of London for further promoting by the authority of experiments the sciences of natural things and of useful arts', and the Society lived up to its title. There was no separation of science and technology, and members were invited to bring for discussion any phenomena they thought would interest the Society. In consequence the most diverse subjects came under review including medical and surgical abnormalities, blood transfusions, experiments with Boyle's air-pump and his pressure-volume relationship, and the natural phenomena of daily life including agricultural operations. King Charles II took a personal interest in the work of the Society and he it was who, having received from Prince Rupert some glass drops of fascinating character obtained in Germany, asked the Society to investigate and report on them. This report, prepared by a committee and read to the Society on 14 August 1661, by Sir Robert Moray, was appended by Merrett to his *Art of Glass*. The chief experiments recorded were repeated from time to time in the early years of the Society to entertain distinguished visitors, and these Prince Rupert's drops have continued to entertain hosts of persons ever since.

The discussions on trades and industries can hardly have failed to include glassmaking because of the numerous forms of apparatus, tubing, and lenses in widespread use by physicians, chemists, physicists, and engineers. Birch in his History refers to four glassworks, Woolwich, the Minories, Rosemary Lane, and Radcliffe, from which supplies were obtained for demonstrations and experimental work at the Royal Society's meetings in 1661 and 1662. There is no clue to the date and circumstances relating to the suggestion of Robert Boyle that Merrett should undertake the translation of Neri's *L'Arte Vetraria*, but it seems highly possible that he received encouragement to undertake the enterprise from his colleagues before the founding of the Society.

Turning now to Merrett's special activities in the Society, Birch records his name more than eighty times during the years 1660–1664, associated with a wide variety of subjects: administrative, surgical, medical, zoological, botanical, chemical, and technological. He served on the Council for the year 1663–1664 and was a member of four of the eight Standing

Committees which by 1664 had been constituted by the Society, namely the History of Trades, the Collecting and Recording of Natural Phenomena, and, as a physician member of the Society, the Anatomical and the Chemical Committees (Birch, Vol. 1, pp. 406–408). At one meeting in 1660 he was elected to serve on a Committee of twelve leading members appointed to consider the erecting of a Library and a Committee to examine the subject of the generation of insects. Special or *ad hoc* Committees to which he was appointed include one to consider the improvement and planting of timber for His Majesty's Navy, one to compile a catalogue of Trades, one to examine a method of making ceruss (basic lead carbonate), one to report on a method of making vitriol, one (a Committee of twelve) to consider schemes for the propagation of cider fruit in England, one (a Committee of thirteen) to consider schemes for the planting of potatoes throughout all parts of England to prevent famine, and one to examine methods of recording experiments and other philosophical matters brought before the Society, their defects, and ways to improve them. Four other major contributions by him are an account of the refining of metals, experiments on freezing, the making, improvement, and keeping of wines, and the preparation and classification of botanical species peculiar to Great Britain.

With the foregoing review of Merrett's qualifications and experience we are better able to assess the value of his contributions to *The Art of Glass* and I shall set out the character of these contributions under four headings:

As translator

How he acquired his substantial knowledge of Italian in order to cope with so specialized a subject as the technology of glassmaking there is no record in his life story, but between 1636 when he graduated M.B. and 1643 when he obtained his M.D. lay six and a half years' further study. Since the medical schools of Italy were among the foremost in the western world he may well have spent some considerable time in that country, like his friend and patron many years his senior, William Harvey, who studied for several years at the University of Padua. In any case by the middle of the seventeenth century there was much correspondence and intervisitation between scientific workers in Italy and England. Later, as a Fellow of the Royal Society, Merrett was on occasion requested to translate memoirs received from scientific bodies in Florence and elsewhere. His translation of Neri appears to have been accepted as satisfactory by his contemporaries who translated Neri–Merrett into other languages, but his was not free from error. One curious mistake appears on page 40 and is repeated on page 57, where *conterie* is translated as 'counting houses' instead of 'beads'. The English version also contains a number of printer's errors, including the repetition in two instances of the same number for different properties of glass (pp. 214–216). Merrett admits that he had considerable difficulty in rendering into English some of the Italian technical terms associated with parts of the glass furnace structure and the tools employed in shaping the glass. In large measure therefore he carried over these names into the English version, but many of them did not become permanently embodied in the vocabulary of English technology.

As commentator

Drawing on his special knowledge of botany and of chemistry, and his wide reading of history both classical and technological, in the course of the hundred and fifty-four pages of *The Art of Glass* occupied by his 'Observations' he quoted the names of more than one hundred individual authorities, in a total of some 350 references. Among these Battista della Porta appears sixteen times, Libavius thirteen times, Pliny, Caesalpinus, Agricola, Cardanus, Kircher, Birelli, and Ferantes Imperatus at least six times, Boethius de Bood five times, and many others two or three times. The object of this wide survey was, according to Merrett in his preface 'To the ingenuous Reader', to bring together all that was known to any good author about glass, its manufacture and use, It is singular, however, that he makes no reference to Vannoccio Biringuccio, the contemporary of Agricola and, like him, a sixteenth-century authority on mining, minerals, and metals, and who included a section on glass in his book *Pirotechnia*.

Merrett naturally lets himself go particularly about those subjects involving botanical knowledge; the naming of the many plants from which glassmaking ash or glass salts may be derived and, in Book VII, the preparation of the sources of colour for lakes from flowering plants for the purpose of painting on glass.

He wrote at considerable length also on the definition of glass, particularly in the ancient world, and on its origin, discovery, and uses. His account of the legends associated with the name of the Emperor Tiberius on the supposed making of malleable glass is tediously long; and he was apt to display loquacity when writing about a subject of which he was imperfectly informed, as for example the meaning, nature, and sources of zaffre. His 'Observations' on this particular colouring agent received the mild rebuke of Johann Kunckel who, in 1679, in his German version of *The Art of Glass*, said quite candidly that Merrett plainly did not know what zaffre was.

Some parts of his long discourse on the nature of glass have passed into technological history and have not infrequently been quoted. After setting down the axiom that glass is one of the fruits of the fire he proceeded (pp. 214–216) to describe the properties of glass '… whereby any one may easily difference it from all other bodies'.

1   'Tis a concrete of salt and sand or stones.
2   'Tis Artificial.
3   It melts in a strong fire.
3   (*sic*) When melted 'tis tenacious and sticks together.
4   It wasts not nor consumes in the fire.
5   'Tis the last effect of the fire.
6   When melted it cleaves to Iron. &c.
7   'Tis ductile whilst red hot, and fashionable into any form, but not mal-

	leable, and may be blown into a hollowness.
8	Breaks being thin without annealing.
9	'Tis friable when cold, which made our proverb, As britle as glass.
10	'Tis diaphanous either hot or cold.
11	'Tis flexible and hath in threads *motum rectitutionis*.
12	Cold and wet disunites and breaks it, especially if the liquors be saltish, and the glass suddainly heated.
13	It onely receives sculpture, and cutting, from a Diamond or Emery stone.
14	'Tis both coloured and made Diaphanous as pretious stones.
15	*Aqua fortis, Aqua Regis,* and *Mercury,* dissolve it not as they do Metalls.
15	(*sic*) Acid juyces nor any other thing extract either colour, tast, or any other quality from it.
16	It receives polishing.
17	It loseth nor weight, nor substance, with the longest and most frequent use.
18	Gives fusion to other Metalls and softens them.
19	Receives all variety of colours made of Metalls both externally and internally, and therefore more fit for Painting than any other thing.
20	'Tis the most plyable and fashionable thing in the world, and best retains the form given.
21	It may be melted but 'twill never be calcined.
22	An open glass fill'd with water in the Summer will gather drops of water on the outside, so far as the water reacheth and a mans breath blown upon will manifestly moisten it.
23	Little balls as big as a Nut fill'd with Mercury, or water, or any liquor, and thrown into the fire, as also drops of green glass broken fly assunder with a very loud & most sharp noise.
24	Wine Beer nor other liquors will make them musty, nor change their colour nor rust them.
25	It may be cemented as Stones and Metals.
26	A drinking Glass fill'd, in part with water (Being rub'd on the brim with the finger witted) yields Musical notes, higher or lower, according as 'tis

more or less full, and makes the liquor frisk and leap.

On the basis of his extensive chemical knowledge, he was able to comment on the preparation of a number of raw materials for glassmaking and on the calcining of lead and other metals. Finally his acquaintance with the writings of the chemists of the sixteenth and early seventeenth centuries, particularly of the former, enabled him to supplement a number of the recipes of Neri for making coloured glasses, pastes, and enamels, and to comment usefully on the preparation of *aqua fortis* and other reagents involved in the preparation of some of the glassmaking materials.

As commentator on the existing state of glass manufacture in England at the middle of the seventeenth century

Merrett was the first to make known the types of glass in course of manufacture in England, the raw materials employed, the sources from which they were derived and their treatment before use, the furnaces in which the preparation of the glass in its different stages was carried out, the kinds of clay used for making the glass-melting pots and the sizes employed for different purposes, the arrangements for gathering the molten glass and for shaping it as required, and finally the means utilized for annealing the objects made.

Taking these items briefly in turn, first of all we learn that by the middle of the seventeenth century progress had been such that manufacture was carried out almost wholly by English workmen and only a comparatively small body of Italian workers remained. Products included window glass, bottle glass, drinking-vessels of various kinds and general domestic glass. and a considerable quantity of glass tubing and special apparatus for scientific purposes. The advance of astronomy had led to the development of a small lens-grinding industry, and Merrett particularly commends Sir Paul Neile, an early Fellow of the Royal Society who had acquired a great reputation for his lenses.

In general three types of glass were made, one which imitated the

Italian crystal glass, an intermediate variety of common white glass, and finally green glass from which bottles and containers, window glass, and probably the stronger scientific apparatus such as stills, retorts, and flasks was made, although some may have been prepared from common white glass. One cannot but be impressed by the hard use to which some of these pieces of apparatus were subjected, in particular the large stills and other vessels which in some operations were liable to be heated continuously for several days on a sand-bath over a coal-fired furnace; their strength must have been substantial.

Astone Gasparetto* has recently written on the considerable manufacture of vessels and apparatus for medical and chemical purposes in Italy in the sixteenth and seventeenth centuries.† In the seventeenth century the records of the early years of the Royal Society also testify to an extensive use of glass vessels in the experiments designed either by the Curator of Experiments or by Fellows who wished to demonstrate experiments they had already carried out.

Of British glass factories Merrett does not appear to have known any other than those situated in London, but he must have known some of these fairly intimately to have been allowed to visit them and become thoroughly acquainted with their methods, apparatus, and tools. He does in fact report that he had discussed problems with them when endeavouring to find English equivalents for Italian terms for the types and parts of the furnaces, the glassworks operators, the tools with which the glass was shaped, and the annealing oven. We might assume that he would know the glassworks at Woolwich and others at the Minories, at Radcliffe, and at Vauxhall; the names of these appear in the early records of the Royal Society as places from which glass apparatus, all glass tubing, and all glass balls or drops were obtained.

No manufacturer of lead glass in this country was known to Merrett. In his "Observations" on Chapter 61, pp. 315–316, he says definitely that 'Glass of Lead', *'tis a thing unpractised by our Furnaces ...*' and then adds.

> *... And could this glass be made as tough as that of* Crystalline *'twould far surpass it in the glory and beauty of its colours ...*

He then proceeds to write quite a lot more about the calcination of lead and the liability of lead as normally calcined to pass through the bottom of the pot. As already pointed out, Merrett does not appear to have been acquainted with glassmaking centres other than London, and it was in London that George Ravenscroft carried out his experiments on the development of a lead crystal glass under the encouragement of the Worshipful Company of Glass Sellers, which had received its Charter in 1664. It can hardly be doubted that Merrett's translation of Neri must have been known to Ravenscroft and his patrons, and may well have assisted in the development of the new lead crystal glass of 1675.

The raw materials for glassmaking and some features of the preparation of the batch are disclosed by Merrett. The London factories were stated to have derived a soft white sand* from Maidstone for the better quality glasses and a harder sand from Woolwich for the green glasses. The use of flints as a source of substantially purer material than sand was mentioned by Merrett (pp. 259–260) but he states that the glassmakers regarded the long slow process of preparing it as making it far too expensive.

In regard to the glassmaking ash one presumes that, for crystal glass, purification was carried out in the manner directed by Neri, but Merrett states that vendors of ash for green glasses journeyed about the country

---

*   *Glastech. Ber.*, V International Glass Congress, special issue, **32K** (1959), VIII, 39–49.

†   Various types of medical glass apparatus were actually in use as early as the first century A.D. Specimens of them were excavated from Pompeii, which was overwhelmed by volcanic eruption and buried in AD. 79, and are now to be seen in the museum on the site.

buying ash from a variety of sources, so that the quality was liable to vary. Merrett also comments on Neri's insistence on grinding the batch mixture, especially for colours, in a mortar following pounding, grinding, and sifting, saying that the mixture was now being prepared by grinding in a horse-drawn mill. These mills were used to grind ashes when in hard lumps, clay salts, manganese, zaffre, and cullet, the millstones being of hard marble seven to eight feet in diameter and nine to ten inches thick. 'This grinding', he said, 'dispatcheth more in one day than 20 men can do in a Mortar'. It must be regarded as one of the early examples of mechanization in glassworks.

As commentator on furnaces

With regard to fuels and furnaces, the British furnaces were now distinguished, and had been so since the beginning of the century, by the use of coal instead of the wood still extensively used on the continent of Europe.

Furnace construction. Merrett refers to three kinds of furnace which were similar to those described by Agricola in the middle of the sixteenth century except that the green glass furnaces of Merrett's time must have been greatly improved in efficiency and temperature. The first type, the calcar, and its use have been described on p. 205. The second was used for melting the crystal glass and took the form of a three-storied beehive-shaped struc-

* Throughout the counties of Kent, Surrey, and Sussex are a number of deposits of sand which were drawn on certainly as early as the thirteenth century by the small glass factories established in the forests of the southern counties and started up probably by glass-makers from Normandy. During the recent world war some of these sources, at Reigate and at Fairlight, near Hastings, were drawn on extensively, and since the war the application of modern chemical treatment processes has led to the output of a very good quality sand from Redhill.

ture separated into segments by vertical ribs which both added strength to the structure and provided boundaries within which workmen, singly or in pairs, confined their operations. On the bottom floor of this furnace was the hearth: from the fire the flame ascended through a central hole in the roof of this chamber which was the floor of the second, or melting, chamber. The flame reverberated against the crown or roof of the second floor down over the pots, but part of its heat was carried through a small circular hole in the crown of the second chamber into a space known as the tower. From the tower led a tunnel forming an annealing oven along which the articles to be annealed were moved stage by stage. Merrett gave this contrivance the name Leer, described by J. Kunckel as a specifically English word.? Thirdly, Merrett says that the green-glass factories were distinguished from those manufacturing crystal glass by using only one type of furnace. This was square (rectangular[†]), having four or more pots on each side, fritt being prepared in one end, and in the other pots pre-heated, the hot gases passing into the ends from the melting furnaces. Apparently the hot gases could also be taken out through a tower in the arch.

As for the materials for construction of the furnaces, it appears that the furnaces for crystal glass were not expected to attain so high a temperature as those melting green glass. For the latter a hard stone was essential which was obtained, he says, from Newcastle, and was of a hard sandy character. Outside these walls of silica stone* was a protection of bricks to ensure a high temperature, and the workmen, according to Merrett, stated that the heat of green-glass furnaces was '... twice as strong as that in the other Glass-furnaces'.

The pots. From whatever source it was obtained the clay was dried and ground before being wetted up. Two types of fireclay are referred to by Merrett, the one obtained from Nonsuch in Sussex, a variety said by him to have been used for tobacco-pipes, the other from Worcestershire which we may presume to have been Stourbridge clay. The latter was needed for pots fired at the higher temperature required for green glass, and he mentions that this Worcestershire clay, or a mixture with Nonsuch, seemed to find general recognition and approval. In all the references to pots they appear to have been open. Two sizes were employed; the larger, in the crystal glass furnaces, holding three or four hundredweight, was two feet deep, twenty inches broad at the top, but considerably narrower at the bottom. In thickness they were nearly two inches at the bottom, tapering to about an inch at the top. The second size of pot was very much smaller; called a piling pot because it stood above and on one side of the large pot, it contained not more than fifty pounds and was used to melt the coloured glasses with which to decorate the crystal glass body.

We learn from Neri that in melting coloured glasses, enamels, and artificial gems the quantity melted and therefore also the size of the pot varied greatly. The lead glasses were often melted in forty- to sixty-pound lots, the enamels in lots of from four to six pounds. while the heavy lead glass gems based on rock crystal were melted in a potter's oven in quantities of ounces only.

The Neri–Merrett *Art of Glass* carries no illustrations, but this brief account of contemporary furnaces and procedures can appropriately be concluded by reproducing illustrations of three taken from the first Latin translation. In this translation Andreas Frisius correlates the descriptions of furnaces and tools by Merrett in his 'Observations' of 1662 with those in use in Amsterdam in 1668. The drawing, Figure 2, of a three-tier furnace for melting crystal glass (similar to the one reproduced by Agricola in his *De re metallica* in 1556) is probably familiar to many readers. Its development about a century later is shown in Figure 3. Three features at least

---

† See my paper entitled 'That curious word lehr'. *J. Soc. Glass Tech.*, 1949, **33**, 278.

* During the first twenty years of the present century Penshaw stone from County Durham was much in use even for tank furnaces where it was in direct contact with the glass.

are to be noted, first the convenient compartment-like working places at the furnace for the glassmaker and his assistant, the glassmaker's chair now appearing for the first time, and the glassmaker's lehr which took the form of a tunnel or nearly level shaft along which the glass objects were slowly moved from hot to cold. Figure 4 includes a further drawing of furnace and lehr and some of the glassmaking tools described by Merrett. Figures 4 and 5 are the earliest known illustrations of glassmakers' tools systematically set out with descriptions.

The influence of Neri–Merrett's 'The Art of Glass'

The publication of Neri's original treatise in 1612 does not appear to have made any great stir. It is rather a coincidence, therefore, that just prior to the publication of Merrett's English translation nearly fifty years after Neri's book, the second impression of the original should appear in Florence, and it is tempting to speculate on the coincidence. Soon after Merrett's publication, however, other Italian editions began to appear, the second in Venice in 1663, a third also in Venice in 1678, and a fourth in Milan in 1817.

The story of the widespread interest which rapidly grew up in western Europe immediately following the publication by Merrett of *The Art of Glass* can rapidly be absorbed by the following summary which I have compiled in the main from *Duncan's Bibliography of Glass* (1961).

Editions of Neri, of Neri–Merrett, and of Neri–Merrett–Kunckel (compiled mainly from Duncan's 'Bibliography of Glass'*)*

I. Neri's L'Arte Vetraria

1612. Italian. Florence.
1661. Second impression. Florence.
1663. Second edition. Venice.
1678. Third edition. Venice.
1817. Fourth edition. Milan.

II. Neri–Merrett: The Art of Glass

Figure 3: Title page of first Latin edition.

1662. The Art of Glass. Merrett's translation of Neri with Observations. English. London.

1826. The Art of Glass. English. Merrett's translation of Neri only, edited by Sir Thomas Phillips. Bart.

1668. *De Arte Vitraria*, the first Latin translation of *The Art of Glass*, by Andreas Frisius, Amsterdam.

1669. *De Arte Vitraria*. Second impression.

1681. *De Arte Vitraria*. Latin, by Andreas Frisius, second edition.
1686. *De Arte Vitraria*. Latin, by Andreas Frisius, third edition.
1678. First German translation by F. Geissler.
1679. *Ars Vitraria Experimentalis, oder Vollkommende Glasmacher-Kunst*. German by J. Kunckel. Frankfurt & Leipzig.
1699. *The Art of Glass*. English Edition of Haudicquer de Blancourt's *De l'art de la Verrerie*. Paris 1697.
1776. Spanish edition, *Arte de Videria*. Dom Miguel Geronimo Suarez y Nunez.

III. Neri–Merrett–Kunckel
1679. German translation of Neri–Merrett, with Kunckel's additions.
1689. Second edition. Frankfurt & Leipzig.
1743. Third edition. Nurnberg.
1756. Fourth edition. Nurnberg.
1752. French translation by d'Holbach, to which d'Holbach adds comments. Paris.

The first Latin edition in which both Neri's text and the 'Observations' of Merrett were reproduced was that by Andreas Frisius at Amsterdam in 1668. There followed three impressions of this work and two further editions, and its author had a number of commendatory remarks to make about Merrett, referring to him as 'the illustrious Dr. Christopher Merrett, Fellow of the Royal Society'.

The first appearance of Neri in the German language was in a translation by Friedrich Geissler in 1678 but this was overshadowed and soon forgotten because a year later Johann Kunckel von Lowenstern published his *Ars Vitraria Experimentalis, oder Vollkommende Glasmacher-Kunst*. In the second half of the seventeenth century he was not only one of the foremost chemists in Germany but was a member of a family several generations of which had traditions as glass-makers. After a wandering career occupied largely by alchemy and important experimental chemistry he became in 1678 a servant of the Grand Duke Frederick II and was

Figure 4: Furnace, lehr crucibles and tools.

attached to Royal glassmaking establishments from 1678 to 1693, first at the crystal glass factory at Drewitz, then subsequently at factories near Potsdam built for experimental purposes. Among these was one built in 1685 on Pfaueninsel for his experiments on ruby glass. Kunckel translated Neri's book with Merrett's 'Observations' and then added a section comprised of valuable comments of his own.* Ten years later, in 1689, he published a second edition containing additional material, whilst after his death the third edition in 1743 and a fourth in 1756 continued to carry forward the valuable work begun by Merrett.

In France in 1752 there was published a version translated by Baron d'Holbach which embodied the texts of Neri, Merrett, and Kunckel with several miscellaneous additions of his own. More than fifty years earlier the influence of Neri and Merrett on French glass-making had been brought to bear by the appearance of Haudicquer de Blancourt's *De l'art de la Verrerie*, published in Paris in 1697. This book has a lengthy introduction on the various branches of the French glass industry and its proprietors, and the

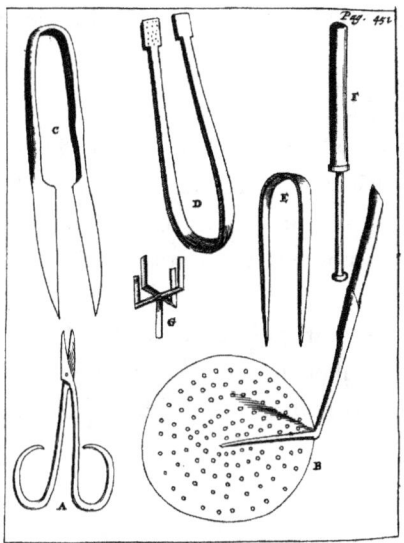

Figure 5: Glassmaker's tools reproduced in the 1668 Latin edition.

claims to original discovery by the writer are not marked by modesty.

Two years later an English translation of this work was published in London and the text follows very closely that of Merrett, the chapter headings in many cases being practically identical. Even in cases where the introductory matter in any chapter varied slightly from that of Neri, so soon as the description of the process begins the text becomes practically identical with Merrett's translation.

Right down to the end of the eighteenth century at least parts of Merrett's translation of Neri continued to be of service. In illustration of this fact, the recipe for making white and crystal glass recommended by the author of the article 'Glass' on page 769 of the third edition of the Encyclopaedia Britannica (1797) is as follows:

> To make crystal glass, take of the whitest tarso, pounded small and serced as fine as flour, 200 pounds: of the salt of polverine 130 pounds: mix them

together, and put them into the furnace called the calcar, first heating it
...

The instructions so given follow closely those set out on page 8 and other early pages in Book I of Neri–Merrett's *The Art of Glass*.

To sum up this review of the influence of Neri–Merrett's book, it can be claimed as well within the bounds of possibility that it had a bearing on the invention by George Ravenscroft in 1675 of English lead crystal glass.

What is certain is that its publication in 1662 set in motion a wave of interest among glassmakers in several countries of western Europe and provided a fund of information from which all could draw, thus forging an international bond.

Acknowledgements

To Professor Douglas McKie and Professor E. N. da C. Andrade, F.R.S.. who contributed so attractively to the Tercentenary accounts in 1960 of the rise and development of the Royal Society, I am indebted for a number of suggestions at the outset of my quest. Mr. I. Kaye, Librarian of the Royal Society, has also been very helpful. My special thanks are due to Miss Margaret Garrett. B.Sc., F.L.S., Assistant Librarian of the Royal College of Medicine, for spending long hours in searching many records in the library of that Institution and in making long abstracts, only a fraction of which I have been able to utilize.

Editor's notes. Dr Robert Brill of the Corning Museum of Glass made the valuable suggestion to include this introductory essay. Turner's text has been slightly modified because it referred to one more figure (showing Merrett's signature) not reproduced here.

---

*   See Johann Kunckel (1630–1703) by H. Maurach, *Abhandlungen und Berichte*, Deutsches Museum, 1933.

THE ART OF GLASS,
by
Antonio Neri
translated by
Christopher Merrett

Imprimatur:

Ex Æd. Sab.
10 Sept.
1662

*G. Stradling Rever. in Christo
pat. Gilb. Episc. Lond.
Sacell. Domest.*

THE
Art of Glass,
WHEREIN
Are shown the wayes to make and colour Glass, Pastes, Enamels, Lakes, and other Curiosities.

Written in *Italian* by *Antonio Neri*, and translated into *English*, with some Observations on the Author.

Whereunto is added an account of the Glass Drops, made by the Royal Society, meeting at *Gresham College*

LONDON
Printed by *A. W.* for *Octavian Pulleyn*, at the Sign of the *Rose* in St. *Pauls* Church-yard. *MDCLXII.*

To the most Illustrious and Excellent Lord *Don Antonio Medici*

Antonio Neri

Having taken much pains for many years from my youth about the Art of Glass, and having experimented therein, many true and marvellous conclusions, I have compiled a Treatise of them, with as much clearness as I could, to the end to publish it to the world, to please and delight (as much as in me lay) men understanding in that profession, having found out many things by my own invention and some others tried by able men, and found most true. I will make manifest those hidden Mysteries, for the reasons abovesaid. If I do attain this my intention it shall occasion me hereafter to be encouraged to publish the rest of my Labours about other Chymical and Physical matters, having likewise in both experimented, many most profitable, credible and admirable Conclusions, for no other reason, but to understand them truly. I judge that I ought not to dedicate this Book to any other but your Illustrious Excellence, who have been always my singular Protector, as also, because you are understanding of this, and of whatsoever Noble

The Epistle Dedicatory

and Precious knowledge, being exercised continually in all these Arts, which are required in a true and generous Prince; I beseech you then to accept, if not the work, yet my devout mind towards your great merit and verture of your most Illustrious Excellence, for whom I pray to God to <u>pour</u> on you all happiness.

From Florence, 6 Jan. 1611.

To the Curious Reader.

THere is no doubt that Glass is one of the true fruits of the Art of fire, since that it is very much like to all sorts of minerals and midle minerals although it be a compound made by Art. It hath fusion in the fire, and permanencie in it; likewise as the perfect and shining Metall of Gold it is refined, and burnished, and made beautiful in the fire. It is manifest that it's use in drinking vessels, and other things profitable for mans service, is much more gentile, graceful, and noble than any Metall or whatsoever stone fit to make such works, and which besides the easiness and little charge wherewith it is made, may be wrought in all places; it is more delightful polite and sightly, than any other material at this day known to the world It is a thing profitable in the service of the Art of distilling, and Spagyrical, not to say necessary to prepare Medicines for man, which would be impossible

To the Curious Reader.

to be made without the means of Glass, so that herewith are made so many sorts of Instruments, and Vessels, as Bodies, Heads, Receivers, Pelicans, Lutes, Retorts, Athenors, Serpentines, Vials, Cruces, square and round Vessels, Philosophical Eggs, Globes, and infinite other sorts of Vessels, which every day are invented to compose and make Elixars, Arcana, Quintessences, Salts, Sulphurs, Vitriols, Mercuries, Tinctures, separation of Elements, all Metalline things, and many others, which every day are found out; and besides there are made others for Aquafortis, *and* Aqua-Regia, *so necessary for Refiners, and Masters of Princes Mints, to Refine Gold and Silver, and to bring them to their perfection; indeed so many things profitable for mans use are made, that seem impossible to be made with out the use of it: and the great Providence of God, is as well known, by this, as in every other thing, who hath made the matter of which Glass is compounded (a thing so needful and profitable to man) so abounding in every place and Region, which with much ease may be every where made. Glass is also a great ornament to the Churches of God, for herewith (besides many other things) are made so many beautiful Glass vessels*

To the Curious Reader.

adorn'd with fair Pictures, wherein the Metalline colours are in such sort advanced, and so lively, that they seem to be so many Oriental Gems, and in the Glass Furnaces, the Glass is coloured with so many colours, with so much beauty and perfection, that it seems no material on the earth can be found like it. The invention of Glass (if it may be like credited) is most antient, for the holy Scripture in the Book of Job, *Chap. 28.* saith, Gold and Glass shall not be equal to it, *&c. which gives clear testimony that Glass was antiently invented, for Saint* Hierom *saith, that* Job *descended from Abraham, and was the son of* Zanech, *who descended from* Esau, *and so was, the fifth from Abraham himself; some will, and perhaps with some reason,* say *that the invention of Glass was found out by the* Alchymists; *for they desiring to Imitate Jewels, found out Glass; a thing perhaps not far from truth; for as I shew clearly in the fifth Book of the present work, the manner of imitating all Jewels, in which way is seen the vitrification of stones which of them selves will never be melted nor vitrified.* Pliny *saith, that Glass was found by chance in Syria, at the mouth of the river Bellus, by certain Merchants driven thither by the fortune of the*

To the Curious Reader.

Sea, and constrained to abide there and to dress their provisions, by making fire upon the ground, where was great store of this sort of herb which many call Kali, *the ashes whereof make* Barillia, *and* Rochetta; *This herb burned with fire, and therewith the ashes & Salt being united with sand or stones frit to be vitrified is made Glass. A thing that inlightens mans understanding with the means, and manner of making not onely Glass, but Crystall and Crystalline, and so many other beautiful things which are made thereof. Many assert that in the time of* Tiberius *the Emperour was invented the way of making Glass malleable, a thing afterwards lost, and to this day wholy unknown; for if such a thing were now known without doubt it would be more esteemed for it's beauty, and incorruptibility, than Silver and Gold; since from Glass there ariseth neither rust, nor tast, nor smell, nor any other quality; Moreover it brings to man great profit, In the use of prospective Glasses and Spheres.*

And although one of them may be made of natural Crystall, called that of the mountain, and the other with the mixture, called Steel, a composition made of Brass and Tin, notwithstanding, in both, Glass is more profitable and

To the Curious Reader.

of less charge, and more beautiful and of greater efficiency: especially in Sphears, which besides the difficulty and expences in making them, they present not to the life as Glass doth, and which is worse, in a short time they grow pale, not representing any thing. Wherefore for these and many other reasons you may well conclude, that Glass is one of the most Noble things which man hath at this day, for his use upon the earth. I having laboured a long time in the Art of Glass, and therein seen many things, I was moved to make known to the World a part of that which I had seen and wrought therein. And although the manner of making Salt, Lees, and Frittaes, is known to many, yet notwithstanding it seemed to me that this matter requires to be handled (as I do) clearly, and distinctly, with some Observations and diligence, which if well considered will not be judged, altogether unprofitable, but perhaps necessary and known to few: besides in my particular way of extracting Salts, to make a most noble Crystall, that if the workman shall be diligent in making it, as I do publish and teach it, with clear demonstrations he shall do a thing as beautiful, and noble, as happily is made in these days, or can be done any other way; and in this thing, in every other,

To the Curious Reader.

matter that I treat of in this present work, the diligent and Curious operator shall find, that I have wrote and shown truth, not told me, or perswaded me by any person whatsoever, but wrought and experimented many times with my own hands, I having always had this aim to write and speak the truth. And if any one trying my receits, and manner of making Colours, Paste, and Tinctures, doth not speed to do so much as I write thereof, let him not be amazed thereat, nor believe that I have writ untruths, but let him think that he hath erred in something, and especially they which have never handled such things; For it is impossible that they at the first time should be masters: therefore let them repeat the work, which they shall always make better, and at the last perfect as I describe it. I warn them in particular to have consideration in colours whose certain and determinate dose cannot be given: but with experience and practice one must learn, and with eye and judgment know when Glass is sufficiently coloured, conformable to the work, for which it ought to serve, and in Paste made in imitation of Jewels conformable to the size, whereof they will make them, Observing that those which are to be set in Gold, with Foyls, as in Rings, or other where, must always be clear, and

To the Curious Reader.

of a lighter colour. But those that are set in Gold to stand hanging in the air, as Pendants, and the like, must be of a deeper colour, all which things it is impossible to teach, but all is left to the judgement of the Curious operator. Observe likewise, and with diligence that the materials and colours be well prepared, and well ground, and that he who will make an exquisite work, may be the securer let him prepare, & make all the colours himself as I teach, for so he shall be sure that his work must happily succeed. The fire in this Art is of notable importance, as that which makes every thing perfect, and without which nothing can be done: Wherefore consideration is to be had in making it in proportion, and particularly with hard, and dry-wood, taking heed of it's smoak, which always hurteth, and endamageth it, especially in furnaces, where the vessels and pots stand open, and the Glass will then receive imperfection, and notable foulness. Moreover, I say that if the operator shall be diligent, and shall do like a diligent and practised person, and shall work punctually as I have set down, he shall find truth in the present work, and that I have onely published, and set out to the world as much as I have tried and experimented. And if I find my pains acceptable to the world, as I hope I shall

To the Curious Reader

be incouraged perhaps to publish my other labors wrought for many years in divers parts of the world in the Chymical and Spagyrical Arts, than which I think there is no greater thing in nature for mans service, known and perfect in ancient times; which made men expert in it to be held for Gods, which then were held and reputed for such. I will not enlarge myself any farther, because I have in the work set down every particular, so clear, and distinct. I rest secure, that he which will not err wilfully, it is impossible he should do so having thereof once made experience and practice. Therefore let all be taken of me in good part, as I have candidly made this present work, first, to the glory of God, and then to the just benefit and profit of all.

To the Honourable,
And true Promoter
of all solid Learning,
ROBERT BOYLE, *Esq;*

SIR,
This Treatise challengeth the inscription of Your name for many reasons. The Author of it Dedicated this piece to a Person of Honour, and eminent parts, both which concurr in you, and herein I thought fit to follow his Foot-steps. Then Your ability to judge of the piece, being for the most part Chymical, wherein You have shewed the world not onely Your great progress singular knowledge, but have also taught it the true use of that most beneficial Art, as to the improvement of Reason and Philosophie. Most Writers therein delivering onely a farrago of processes and unintelligible Enigm'as. But You have chalked out the way of solid reasoning upon whatsoever occurrs to observation in such experiments. Next, you were the principal cause

that this Book is made publick, by proposing and urging my undertaking of it, till it came to a command from that most Noble Society, and serious indagators of Nature, meeting at Gresham College, *whose desire I neither could nor ought to decline. Though their, and your choice might have been much more happy, there being many of that company far more adapted for this undertaking than myself. Besides, I doubt not but You will much promote by Your practice the Art it's self, there being scarcely any thing contained in it, but You have already judiciously had experience in. Not, because this Translation will any whit avail You (since Your skill in the native Language is sufficiently known to all that have the honour to be acquainted with you) but may be compendious to You for such as You shall employ in these operations. Furthermore I have herein also satisfied Your vast desire of communicating knowledge to others, who though intelligent of the Language could not procure Copies in the Original; And lastly the candor of your genius no less than that of your intellectuals ready to excuse the errours, and slips whatsoever of,*

*Sir, Your most humble
and most regard-
ful Servant
C.M.*

To the ingenuous Reader.

Courteous Reader,
I Am to advertise thee of some things, concerning the Translation of this Book. You may take notice that I had first Translated it word for word, but finding that the Author had thorewout the whole, so often repeated the same thing, by advice of some ingenuous persons, I left out those repetitions, and have either before the Books given a general account of these repetitions, or else have referr'd you to a former process where the latter hath reiterated the same, and for the most part in the very same words, yet so that I have omitted nothing material in the Author: For what need is there to say as often as Manganese *is boil'd with the metall, that you must do thus and thus, lest it run into the fire, &c? or to repeat the same process, and rules in each new colour for Pastes or Glass*

To the Ingenuous Reader.

of Lead? Though you may find some needless repetitions too, in this Translation not omitted. I confess these reiterations caus'd a nausea in my self, and believe they would in thee, and therefore I passed them over. Then observe that there being many words peculiar to this Art, I was compell'd to have recourse to the workmen, and for such things, and materials not used nor known here, to take them upon trust from such workmen as have wrought at Muran *and other parts of* Italy. *As for other things I have carefully surveid them my self. Now for the observations I have been more large, especially in a business, wherein so little hath been said, and therefore have delivered whatsoever is material that I have met with in any good Author concerning whatsoever* Neri *treats of, that thou might'st have together all that is substantially written upon this unusual subject and have supplied some things defective in our Author, or very fit to be known to Curious persons. Lastly, I doubt not but our workmen in this Art will be much advantaged by this publication, who have within these twenty years last past much improved themselves (to their own great reputation, and the credit of our nation) insomuch that few foreiners of that profession are now left amongst us. And*

To the Ingenious Reader.

this I the rather say, because an eminent workman, now a Master, told me the most of the skill he had was gain'd by this true and excellent book (they were his own words,) And therefore I doubt not but 'twill give some light and advantage to our Countrey-men of that profession, which was my principal aim. And lastly for the exotick words you'l meet with in Reading this Book they are now current with us, or else expounded in my observations.

Fruere & utere.
C.M.

To avoid our Authors Re-petitions, Observe

1. ALL *the fires must be made with dry and hard Wood.*
2. *When the Glass is coloured before you work it, mix the colours well (which otherwise sink to the bottom of the pot) with the metall that the Glass may be coloured through-out. This must be observed all the time you work the Glass into any vessels.*
3. *The sign that Brass or Copper are well calcin'd is that they being put into the metall, make it swell and suddenly rise, if they be calcin'd too much or too little, those signs are wanting and Glass made thereof will be Black and foul.*
4. Manganese *consumes the natural greenness of Glass.*
5. *Copper, Brass, Lead, Iron, and all compositions of them as also* Manganese, *must be put into the metall, but a little at a time, and at convenient distances, and the pot must be large and not filled too full, because they all swell well and rise much, and so are apt to run over into the fire to your loss.*

To extract the Salt of Polverine, Rochetta, *and* Barillia *wherewith crystall Fritt, called* Bollito *is made.*

The foundation of the Art of Glass-work, with a new and secret way.

CHAP. I.

Polverine, or *Rochetta*, which comes from the Levant and Syria, is the ashes of a certain herb growing there in abundance: there is no doubt but that it makes a far whiter salt than *Barillia* of *Spain*, and therefore when you would make a Crystall very perfect and beautifull, make it of salt extracted from *Polverine* or *Rochetta* of the *Levant*. For though *Barillia* yield more salt, yet Crystall made therewith alwaies inclines to a blewness, and hath not that whiteness and fairness as that made of Polverine hath.

  The way often by me practised to extract the salt perfectly from both of them, is this which follows.

  Powder these ashes, and sift them with a fine sieve, that the small pieces go not thorow, but only the ashes; the finer the sieve, the more salt is extracted. In buying of either of these ashes, observe that they abound in salt; this is known by touching them with the tongue, and tasting what salt they contain: but the safest way of all is, to make an essay of them in a melting-pot, and to see whether they bear much sand, or *Tarso* a thing common in this Art and which the Conciators very well know.

  Set up brass coppers with their furnaces like those of the Dyers, greater or lesser, according as you have occasion to make a greater or lesser quantity of salt:

fill these coppers with fair and clear water and make a fire with dry wood, and when the water boyleth well, put in the sifted *Polverine* in just quantity and proportion to the water, continue the fire and boyling till a third part of the water be consumed alwaies mixing them at the bottom with a scummer, that the *Polverine* may be in-corporated with the water and all its salt extracted; then fill the coppers with new water, and boyl it till half be consumed, and then you have a lee impregnated with salt. But that you may have salt in greater quantity, and whiter, put into the coppers when they boyl, before the *Polverine* is put in, about 12 pound to a copper of *Tartar* of red wine, calcined, only to a black colour, dissolve it well in the boyling water, mingling it with a scummer then put in the *Polverine*. This way of *Tartar* is a secret, and makes more, and whiter salt, and a more beautiful Crystall. When two thirds of the water is evaporated, and the lee well impregnated with salt, slacken the fire; under the copper and have in order many earthen pans, filled with common water for six daies, that they may imbibe less lee and salt, and then with great brass ladles, take the lee out of the copper, and put it into the said pans, take out also the ashes from the Copper, and put them all in to the same pans, and when they are full, let them stand for ten daies, for in that time the ashes will be all at the bottom, and the lee remain very Clear, then with brass ladles, take gently (that the bottom be not raised, and troubled) the clear lee, and put it into other empty pans, and so let the lee stand two daies, which by the setling of more terrestriety at the bottom, becomes very clear, and limpid, let this be thrice repeated, and you shall have the lee most limpid, and discharged of all terrestriety, wherewith a very fine and perfect salt is made. Let the coppers be filled again, and boyl with the same quantity of *Tartar*, and then the *Polverine* as before; Continue this work till you have materials enough.

    To strain the said lees, and extract the salt, first wash the Coppers well with clear water, then fill them with the said refine and clarified lees, and make them boyl softly, as before, and observe that you fill the coppers with the said lee, till you see it thicken, and shoot its salt; which is wont to be about the beginning of 24 hours, for then in the superficies of the copper, you will begin to see white salt appearing like a spiders web, or white thread, then hold a Scummer full of holes at the bottom of the copper, and the salt will fall upon it, and now and then take it out suffering the lees to run out well off it into the copper, then put the salt into tubs or earthen pans, that the lee may be better drained, the liquor that drains must be saved, and put into the copper, then dry the salt. Continue this work till all the salt be gotten out of the copper: but you must observe, when the salt begins to shoot, to make a gentle and easie fire, for a great fire makes the salt stick to copper, and then the salt becoming strong, alwaies breaks the copper, which thing hath sometimes hapned to me; wherefore observe this chiefly, using great patience and diligence. The salt in the pans, or tubs, being well drained, must be taken, and put into wooden tubs, or vats, the better to dry out all the moysture, which happens in more, or fewer daies, according to the Season in which it is made.

The secret then of making much, and good salt, consists in the *Tartar*, as is before demonstrated. From every three hundred pound of ashes, I usually get from 80 to 90 pound of salt. When the salt is well dryed, beat it grossly, and put it into the Calcar to dry, with a most gentle heat, and with an iron rake it must be broken, and mixed as the Fritt is; when it is well dryed from all its moisture, observing alwaies that the Calcar be not very hot, but temperate, take it out of the Calcar, and pound it well and sift it so, that the greatest pieces which pass thorow, exceed not the bigness of a grain of wheat.

  This salt thus pounded, sifted and dryed, must be kept by it self, in a place free from dust, for to make Fritt of Crystall: the way to make this Fritt is this which follows.

The first Book.

The way to make Fritt for Crystall, otherwise called Bollito.

CHAP. II.

WHen you would make fair, and fully perfect Crystal, see you have the whitest *Tarso* which hath not black veins, nor yellowish like rust in it. At *Muran* they use the pebles from *Tesino*, a stone abounding in that River. *Tarso* then is a kind of hard, and most white marble found in *Tuscany*, at the foot of the *Verucola* of *Pisa*, at *Seraveza*, and at the *Massa* of *Carara*, and in the River *Arnus*, above and below *Florence*, and it is also well known in other places. Note, that those stones which strike fire with a steel, are fit to vitrifie, and to make glass and those which strike not fire with a steel, will never vitrifie, which serves for advice to know the stones that may be transmuted, from those that will not be transmuted into glass.

Take then of the best *Tarso*, pounded

small, serced as fine as flower, 200 pound; also salt of *Polverine* pounded, and sifted also about 130 pound, mix them together, then put them into the Calcar which at first must be well heated, for if they be put into the Calcar when it is cold Fritt will never be made of them. At first for an hour, make a temperate fire and alwaies mix the Fritt with the rake, for this is a thing of great importance and you must alwaies do thus for 5 hours still continuing a strong fire.

    The Calcar is a kind of calcining furnace, the rake is a very long instrument of iron, wherewith the Fritt is continually stirred; both these are very well known, and used in all glass furnaces. At the end of 5 hours, take the Fritt out of the Calcar which in that time (having had sufficient fire, and being well stirred) is made and perfected. Then put this Fritt in a dry place on a floor, and cover it well with a cloath, that no dust nor filth may fall upon it: for herein must be used great diligence, if you will have good Crystall. The Fritt

The first Book.   9

thus made, becomes as white as snow from Heaven. When the *Tarso* is lean, you must add somewhat more than ten pound of the salt to the quantity aforesaid. Wherefore let the experienced Conciators alwaies make tryal of the first Fritt, by putting it into a chrysible, which being put into the furnace, if it grow clear, and suddenly, they know whether the Fritt be well prepared, and whether it be soft, or hard, and whether quantity of salt is to be increased, or diminished. This Crystall Fritt must be kept in a dry place, where no moisture is, for from moist places, the Fritt suffers much, the salt will grow moist, and run to water, and the *Tarso* will remain alone, which of it self will never vitrifie: neither is this Fritt to be wetted, as others are. And when it is made, let it stand 3 or 4 months, and it will be much better to put into the pots, and sooner waxes clear. This is the way to make Crystall Fritt, with the dose and circumstances, which I have oft times used.

Another way to extract salt of Polverine, which makes a Crystall as fair and clear as natural Crystall. This was my invention.

CHAP. III.

Take *Polverine* of the *Levant* well serced and put it into great glass bodies luted at the bottom, with ashes, or sand, into the furnaces, filling them at first with common water, give them a temperate fire for some hours in the furnace, and let them stand till half the water be evaporated; the furnace being cold, gently decant off the water into earthen pans glased, putting new water upon the remainder of the *Polverine*, and let it boil (as before) this is to be repeated till the water hath extracted all the salt; which is known, when the water appears to the tast not at all saltish, and to the eye when it is void of colour. Take of these Lees what quantity you will, let them be filtred, and stand

The first Book. 11

in glased pans four or six days to settle, which by this means will leave a great part of their terrestriety, then put them to filtre anew thus will they be purified, and separated from a great part of their terrestriety, then let these Lees be set to evaporate in great glass bodies, luted at the bottom, in furnaces, in ashes, or sand, at a gentle fire, and at last when the stuff is dryed, observe that the fire be very gentle, that the salt be not burned nor wasted.

When the salt is dried, take out the glass bodies and see if they be broke at the bottom, which is wont to happen often, in which case put the said Salt into other good glasses, well luted at the bottom, and fill them at the top with common pure and clean water thus which set in the furnace, in ashes or sand, at a gentle fire, and always evaporate an eighth part of the said water, then, the furnace being cold, empty this water fully impregnated with salt into earthen pans glased, and when the water is setled 24 hours, filtre it with diligence, that the salt may be separated from the rest of the terrestriety and dregs, let this lee be evaporated in glass bodies with a gentle fire, and at last more gentle, that the salt be not burned, put this salt again into glass bodies to be dissolved in common water, in every thing as before repeat this work till the salt yields no more terrestriety, or dregs, then shall you have a pure and perfect salt wherewith a Fritt made with *Tarso* as before, will make a Crystall, which in fairness, whiteness, and cleerness, will excel natural Crystall.

An observation for Gold Yellow, in Crystall.

CHAP. IV

OBserve that when salt of *Tartar* is mixed with salt of *Polverine*, Fritt made of the said salt is not good to make nor can make, a Gold yellow, although it make all other colours. But to make your Gold yellow, you must make Fritt with salt taken from *Polverine* alone, first purified as above, for otherwise this yellow will not arise. Although this process be somewhat laborious, and a small quantity of salt made therewith, yet not withstanding it will make a Crystall worthy of all great Princes, being fit to make all sorts of vessels and works. This was my invention, whereof I have many times made trial with happy success and my great content.

The way to make salt of Fern, which makes a very fair Crystall.

CHAP. V.

IN *Pisa* I made experience of Fern-ashes which groweth in great abundance in *Tuscany*, which herb must be cut from the earth, when it is green, from the end of *May* to the midst of *June*, and in the Moons increasing, when it is near it's opposition with the Sun; for then the said herb is in it's perfection, and will then make more salt, and of a better nature, strength and whiteness, than at other times: for when it is suffered to dry of its self upon the ground it yields little salt, and of little goodness. This herb being thus cut and laid together, soon withereth, then let it be well burned to ashes, with these ashes, and with the rules, observations, and diligence given before for the salt of the *Polverine*, of the *Levant*, is extracted a pure and good salt wherewith I have made Fritt with good and well serced *Tarso*, the which Fritt melts well in the pot, and yieldeth a fair Crystall, and much better than the ordinary Crystall, because it had more strength and bended much better, which the ordinary Crystall doth not, it is drawn into fine threds, such as I caused to be drawn, and to this Fritt may be given a wonderful yellow Gold colour, observing that there be in it no salt of *Tartar*, for neither from this will the Gold yellow arise, and the Gold yellow which is given to this Crystal is much fairer and pleasanter than can be wrought with the Crystall made with the salt of the *Levant Polverine*, and with that Crystall can not be made all sort of works as with the other.

The way to make another Salt which will produce a marvelous and wonderful Crystall.

CHAP. VI.

Let there be made ashes after the manner aforesaid, of the Cods and Stalks of Beans, dried in the summer, when the husbandmen have thrashed and separated the Beans with which ashes, with the rules and pains abovesaid in the salt of *Polverine*, a marvellous salt is extracted, wherewith is made a very noble Fritt with white and well serced *Tarso*, which in pots will make most beautiful Crystall; the same may be done with the ashes of Coleworts, Bramble Berry bush, and also with stalks of Millet, Rushes and fen Reeds, and many other herbs which yield a salt, wherewith (making Frits after the accustomed manner) will be made most beautiful Crystall, as every noble and curious spirit try by experience, for thereby more is learned than by long study.

A salt that will make a very fair Crystall.

CHAP. VII.

TAke the salt of Lime which serves for building, and this Salt purified and mixed with the ordinary Salt of *Polverine* of the *Levant* about two pound to a 100 that is two pound of salt of Lime to 100 pound of the salt of *Polverine* purified and well made (as abovesaid) with this salt so mixed is usually made ordinary Fritt; and is put in the pot to <u>refine the Glass</u> as shall be hereafter declared in the way of making of Crystalline, Crystall, and common glass, and so <u>is made</u> a very fair and beautiful Crystall.

The way to make ordinary Fritt, to wit of Polverine, Rochetta, and Barillia of Spain.

CHAP. VIII.

FRitt is nothing else but a calcination of those materials which make glass, although they may be melted, and make glass without calcination, yet this would succeed with length of time and weariness, and therefore this calcination was invented to calcine the Fritt in the Calcar, which, when it is calcined, and the proportion of the materials is adjusted agreeable to the goodness of the *Barillia*, presently melts in the pot, and wonderfully clarifies.

Fritt made of *Polverine* makes ordinary white glass; Fritt from *Rochetta* of the *Levant* makes the fairest glass called Crystall; *Barillia* of *Spain*, though it be usually fatter than the former, yet it makes not a glass so white and fair as that of the *Levant*, because it always inclines a little to an azure colour.

To make then Fritt, serce the *Polverine* thorow a fine serce, the small pieces which pass not, let them be pounded in stone mortars the like is to be done with the *Rochetta* and *Barillia*, to wit every one by it's self and be sure that they be well pounded, and serced thorow a fine serce for as the common proverb in this art saith; A fine serce, and dry wood, bring honour to the furnace. Now whatsoever the quantity of the *Barillia* be, for example, a 100 pound of it commonly requires, from 80 pound to 90 of *Tarso*, which must be fine beaten, and then finely serced, more or less, according to the goodness of the *Barillia*, and it's fatness, whereof you need not make an essay, how much it holds as is known by art. Then with sand, and especially with that from *Tuscanie* found in the vale of *Arnus*, being much fatter, and having in it more plenty of salt, than *Tarso* hath. There is never added more than 6 or 8 pound to the hundred. Now this sand must be washed from all it's unprofitable terrestriety, and serced, and then this will make a white and good glass for *Tarso* always makes much fairer glass than any sand that is in *Tuscany*. The due quantity

The first Book. 19

of Sand or *Tarso*, being found out, mix and unite them, first well together with *the Barillia* or *Polverine* well sifted, and so put them into the Calcar when it is hot, & at first mix & spread them well in the Calcar with a rake, that they may be well calcined and continue this till they begin to grow into lumps, and come into pieces big as small nuts, The Fritt will be well and perfectly wrought in the space of 5 hours, being stirred all that time, and a sufficient fire continued, and when you would see whether it be well made, take a little of it out, when cold, if it be white, yellowish and light, then 'tis made. The calcining of it more than 5 hours is not amiss; for by how much more it is wrought and calcined, the better it is, and the sooner it melteth in the pot, and by standing a little longer in the Calcar it consumeth and loseth the yellowness and foulness which glass hath in itself, and it becomes more clear and purified. When the Fritt is taken out of the Calcar thus hot, let there be thrown upon it 3 or 4 pails of cold water, then set it under ground, in a moist and cold place, and the filth which arose when the salt was made (as is above said) is wont to be put into the same pans, with the lee from *Polverine*; fill them with common water, having tubs under the pans to receive the water, which by little and little drops thorow the said filth and setlings, and hence comes a very strong pure and clear lee to be kept apart and herewith now and then water the Fritt abovesaid, which being heaped together in a moist place the space of 2 or 3 moneth or more (the longer the better) then the said Fritt grows together in a mass as a stone, and is to be broken with mattocks. Now when it is in the pot it soon melted stupendiously, and maketh glass as White as Crystall. For this Lee leaves upon the Fritt it's salt which worketh this effect. When this Lee is not to be had it must be watered with common water, which although it doth not work this great effect as the said lee, yet it doth well, and maketh it easier for melting. Wherefore Fritt should stand, when made, some moneths; which thus made alway causeth less wood to consumed, and the glass clear and sweeter to work.

To make Crystall in full perfection, the way I always practice.

CHAP. IX.

TAke Crystall Fritt diligently made, set it in pots in the furnace, where there are no pots with colours, for the fumes of metals whereof the greatest part of colours are made, make the Crystall pale and foul but that it may come forth white, shining, and fair, when you put the Fritt into the pots in the furnace, then cast in such a quantity of *Manganese* prepared as is needful, according as the pots are greater or less. For this lieth in the practice of the able and diligent Conciatore, and belongs to his office. The quantity of *Manganese* and of all other colours to be into the Fritt and metals cannot be precisely determined either by weight or measure, but must be wholly left to the eye and judgement, tryal and experience of the Conciatore. To make a fair Crystal, when it is well melted take it from the pots; and cast it into great earthen pans, or clean bowlsful of clean water (for it requireth to be cast into water) to this end that the water may take from it a sort of salt called *Sandever* which hurteth the Crystall, and maketh it obscure and cloudy, and whilst it is working still casteth forth *Sandever*, a thing very foul. Then put it again into a clean pots and cast it into water which is to be repeated as often as is needful, until the Crystal be separated from all this salt, but this is to be left to the practice of the Conciator, then set it to boil 4 or 6 days and let as little Iron be mixed therewith as is possible, for it gives it always a blackish tincture. When it is boiled and clear, see whether it hath enough *Manganese*, and if it be greenish, give it *Manganese* with discretion. Wherefore to make good Crystall put in the *Manganese* by little and little at a time, for it makes the Crystall of a murry colour, which afterward inclines to black taking from it it's splendor; mix the *Manganese*, and let the glass clarifie till it becomes of a clear and shining colour. The property of *Manganese* is, being put in just quantity to take away the foul greasiness which Crystall always hath, and to make a resplendent white, when the Crystall is clear limpid & fair, work it continually into vessels and works that most please you, but not with so great a fire as common glass is wrought with. Be careful that the Irons wherewith you work be clean, and that you put not the necks of the glasses where the Irons touch (for there always remaineth Iron) into the pots of Crystall, for they make it become black. But this glass where Iron rods touch may be put in to make glass for vulgar works.

To make Crystall-Glass, and white, call'd otherwise common glass.

CHAP. X.

FRitt of *Polverine* makes a white and fair common glass, Fritt of *Rochetta* which the fairest glass called Crystall, is between ordinary glass and Crystall. As much *Manganese* prepared must be used in common glass as is in Crystalline; cast the Crystalline or common glass once at least into water, that you may have them fair, clear, and in perfection. Although glass may be made without this casting into water, yet to have it fairer than ordinary, this is necessary to be done, and maybe repeated according to your pleasure, as you would have them more resplendent and fairer, and then you may work them into what vessels you need. And to make them yet whiter, Calcine them that they purifie well and have but few blisters. And above all observe, that if to each of them, by themselves, you put upon the Fritt, the proportion of 10 pound of salt of *Tartar* purified to a 100 weight of Fritt, it makes them without comparison fairer, and more pliable to work than ordinary. The salt of *Tartar* must be put in when the Fritt is made, and then be mixed with *Tarso*, or sand, together with the *Polverine* or *Rochetta* sifted, and then make thereof Fritt as before.

To make Purified salt of Tartar, *for the work abovesaid.*

CHAP. XI.

TAke *Tartar* of red wine in great lumps, and not in powder, Calcine it in earthen pots between live coles till it become black, and all it's unctuosity be burned away, and till it begins to grow white but let it not become white, for then the salt will not be good. Put the said *Tartar* thus Calcined into great earthen pans full of common water heated, as also into earthen pots glased, & make it boil with a gentle soft fire in such sort that a quarter of the water may be exhaled two hours, then take them from the fire and suffer the water to cool, and become clear, which decant off, and it will be a strong lee, then put in more common water into the said pans after the same manner, and upon the remainder of the *Tartar*, and let them boil as before, repeating this until the water become no more saltish, then Filtre those waters impregnated with salt, and put the clean Filtred lee into glass bodies to evaporate in the ashes of the furnace at a gentle heat, and in the bottom there will remain a white salt, dissolve this salt in warm water, let it settle two days, then evaporate it in glass bodies at a gentle heat, and there will remain the bottom a salt much whiter than at the first time, dissolve this salt again, and after two days setling, Filtre and evaporate it in every thing as before. Repeat this manner of dissolving, Filtring, evaporating this salt of *Tartar* four times, which then will be salt much whiter than snow, and purified from the greatest part of it's Terrestriety which salt mixed with *Polverine* or *Rochetta* serced with a dose of *Tarso* or sand will make a Fritt, which in the pot will yield Crystalline and common glass much fairer than that that is made without the addition of this salt of *Tartar*, and although a fair Crystalline may be made without it, yet notwithstanding a much fairer may be made with it.

To prepare Zaffer *which serves for many colours.*

CHAP. XII.

TAke *Zaffer* in gross pieces, put it into earthen pans, let it stand half a day in the furnace & then put it into an Iron ladle to be heated red hot in the furnace, take it thence thus hot and sprinkle it with strong vineger, as soon as 'tis cold grind it fine on a Porphyrie stone, wash it in earthen pans glased with much warm water, always suffering the *Zaffer* to settle to the bottom, then decant it gently off, this will carry away the foulness and Terrestriety from the *Zaffer*, and what is good, and the tincture thereof will remain at the bottom, which prepared and purified will tinge much better than at first, making a limpid and clear tincture, which dry and keep in vessels closed for use.

To prepare Manganese *to colour glass*

CHAP. XIII

Take *Manganese* of *Piemont*, for this is the best of all the *Manganeses* at this day known in the art of glass. At *Venice* there's not always plenty, and at *Moran* none other is used. In *Tuscanie* and *Liguria* there's enough; but that holds much Iron, and makes a black foul colour. That of *Piemont* makes a very fair murry, and at last leaves the glass white, and takes away the greenness and blewness from it. Put this *Manganese* in pieces into Iron ladles and proceed thorowout as in preparing *Zaffer*.

To make Ferretto *of* Spain *which serves to colour glass.*

CHAP. XIV.

TO make Ferretto is nothing but a simple Calcination of Copper, that the metall being opened, may communicate its tincture to glass; which Calcination when it is well made without doubt makes divers and very beautiful colours. This calcination is made many ways, I shall set down two of them, not only easie but of times used by me, with effects very fair, in glass, whereof the first is this that followeth, to wit, Take thin Copper-plates of the bigness of a *Florentine*, and have one or more melting pots of the Goldsmiths, and in the bottom of these pots make a layer of brimstone powdered, then a layer of the said plates, and over them another layer of powdered brimstone, and another Copper-plates, as before, and in this order fill the pot, which is otherwise said to make a SSS*, cover and lute well, and

---

*   *Stratum super stratum* = layer upon layer

dry this pot, and put it into an open wind furnace amidst burning coals, and a strong fire must be given to it for 2 hours, let it cool, and you shall find the copper Calcined, and it will be broke in pieces by the fingers as if it were dry earth, and will be raised into a black and reddish colour. This Copper being beaten small and serced in a fine serce is kept well closed for use.

Another way to make Ferretto *of* Spain.

CHAP. XV.

THis second way of making burnt Copper, though it be more laborious than the first yet it will do it's effects in glass more than ordinary.

The Copper then (instead of making a SSS with Brimstone) must make a SSS* with Vitriol, and then Calcine it, letting it stand three days in the floor of the furnace, near the occhio, then take it out & make another SSS with new Vitriol, keep it in reverberation as before, & this Calcination with Copperas

---

*   Stratum super stratum = layer upon layer

The first Book. 31

must be repeated six times, and then you shall have a most noble burnt Copper, which in colours will work more than ordinary effects.

To make Crocus Ferri, *otherwise called* Crocus Martis, *to colour glass.*

CHAP. XVI.

C*Rocus Martis* is nothing else but a subtilising and Calcination of Iron, by means whereof it's tincture (which is most red in glass) is so opened that it communicateth it's self to glass, & not only manifesteth itself, but makes all other metalline colours (which ordinarily are hidden and dead in glass) appear fair and resplendent; I will set down four ways to make it, and the first is.

Take filings of Iron (if you can have them those of steel are better) mix them well with three parts of powdered brimstone and keep them in a melting pot in a furnace to Calcine, and burn well off all the brimstone, which soon succeeds, let them stand four hours in burning coals then take and powder, and serce them thorow a fine serce and put them in to a Chrysible covered and luted at the top, & set them in the Leer of the furnace neer the occhio or the cavalet 15 days or more which then gains a reddish Peacock-like colour, as if it were purple, this is kept in a close vessel, for the use of glass colours, for it worketh many fair feats.

The second way to make Crocus Martis.

CHAP. XVII.

THis second way of making *Crocus Martis*, with so much ease, ought to be much esteemed of, since the *Crocus* made in this manner makes appear in glass the true red colour of blood, and the manner of making it is thus.

Take filings of iron (steel is better) mix them well in earthen pans with strong vineger, onely sprinkling them so much they may be wet thorowout, spread them in pans, and set them in the sun till they be dry, or in the open air when the sun is cloudy. When dry, powder them, if they be any whit in lumps, sprinkle them with new vinegar, then dry and powder them as before, repeat this work 8 times, then grind and serce them fine, and you have a most fine powder of the colour of brick powdred, which keep in vessels to colour glass.

A third way to make Crocus Martis.

CHAP. XVIII.

THis third way of making *Crocus Martis* is a way by which the deep colour of Iron is made more manifest than may seem credible, and in glass is seen the truth and proof thereof. Sprinkle filings of steel with *Aqua-fortis*, in glased pans, set them in the sun to dry, powder them, wet them again with *Aqua-fortis* and dry them, repeat this several times, and you shall have a red powder, as is said of Crocus made with Brimstone, then powder, serce, & keep it for your use to colour glass.

A fourth way to make Crocus Martis.

CHAP. XIX.

THis is the fourth and last way to make *Crocus Martis*, and perhaps the best of all, though each of the ways shown by me are not onely good and perfect in their operation but necessary also for divers colours necessarily & daily made in glass, & to make this, dissolve in *Aqua fortis* made *Aqua-Regis*, with *Sal Armoniack* (as shall be said in our rules of *Calcidony*) filings of Iron or steel in a glass vessel well closed, keep them so 3 days, & every day stir them well. Observe, when the said water is put upon the filings, that it be done leasurely, & warily, because it riseth much, and endangereth the breaking of the glass, or else all to run out: at the end of 3 days let the water gently evaporated away, and in the bottom will be found a most noble Crocus Martis for the most stupendious tinctures of glasses, which keep for use.

To Calcine Brass called Orpello *or* Tremolante, *which in glass makes a skie colour, and sea green.*

CHAP. XX.

BRass (as it is well known) is Copper, which by *Lapis Calaminaris* becomes of the colour of gold, the which *Lapis Calaminaris,* doth not only colour the Copper, but also incorporating with it increaseth much its weight, the which augmentation gives a colour to glass when it is well Calcined, which is a thing very delightful to see, keeping the medium between a Sea-green and a skie-colour, when the Skie is clear and serene, wherefore be diligent in well Calcining it; to make it punctually, this is the way.

Take Brass, and to save charges, buy that which is in works, and Festoons, cut it in small pieces with a pair of Scisers, then put it into a Chrysible covered and luted at the top in coals on a strong fire. I put it in the burning coals of the furnace where they are stirred, there let it then stand four days in a great, but not in a melting fire, for if it be melted, all the labour is lost, in that time it will be well Calcined, powder it into a most subtil powder, and serce it, and grind it fine upon a porphyrie stone and there will come forth a black powder which spread on tiles, and keep it on the burning coals in the leer, near to the round hole, four days, take from it the ashes that fall upon it, powder, serce, and keep it for use. The sign that it is well Calcined is that if it be put into glass it makes it swell and when it makes not the glass arise and boil well, it is a sign, either that it is not well Calcined, or that it is too much burnt, in which two cases, it makes not the glass boil, neither doth it colour well.

To Calcine the said Brass, after another manner, to make a transparent red, a yellow, and Calcidony.

CHAP. XXI.

TAke Brass and cut it small with sheers, and put it in melting pot, make a SSS. with powdered Brimstone, and set it on kindled coals, put it in the burning coals of the furnace for 24 hours, then powder, serce, and put it covered upon tiles of earth into the furnace, for 12 days to reverberate, then grind, powder, keep it for use.

Sea-Green in glass, a principal colour in the Art.

CHAP. XXII.

SEa-green is one of the principal colours given to glass, and if you would have it fair and to hold at all trials, you must always make it in glass called Artificial Crystal; for in ordinary metall it riseth not fair: and though in Crystalline it ariseth fairer than in common glass, yet in the said Crystal, onely in perfection. Observe, that whenever you would make this colour, you in no wise add any *Manganese* at first, because this being added (although the fire consumes it,) yet it leaves a quality in the glass, which makes the colour black, and gives it great imperfection and foulness. Now to make a fair Sea-green, take of Crystal Fritt, and put it in a pot, not allowing it any *Manganese* at all, and as soon as it is melted and clear it yields a salt which swims at the top like oyl, let this be taken off with Iron ladles

The first Book.

by the Conciators, take it out with much diligence, for what remains thereof, will make a foul and oyly colour, and when the glass is well and perfectly clarified, take a pot of about twenty pound of Crystal, six ounces of Brass prepared as is aforesaid, and to this Brass calcin'd, add a fourth part of *Zaffer* prepared, and let these two powders be well mixed, and put to the said Crystal at three times, but at first this powder will make the metall swell very much, and you must well mix the glass with the long squares. Then let the metall settle, that the colour may be incorporated for 3 hours, then mix them again with the long square, then take a proof thereof, in doing whereof, put in rather too little than too much, for the colours may be easily heightned, which is to be done according to the works for which it is to be employed, for gross tubes for beads[*] require not so deep and full a colour and tubes to make beads of, must not have too light a colour. At the beginning of twenty-four hours (after it hath had the due colour) it may be wrought, and before you work it, mix well the metall from the very bottom of the pot, that the colour may

---

[*]   Merrett has *counting houses*; a rare mistranslation.

be well united, mixed, and spread thorow all the metall, otherwise it settles the bottom, and the metall at the top comes clear. This manner of making Sea-green, I have tried at *Florence* in the year 1602 and made pots for tubes for beads, always of a most fair colour. At *Moran* for the said tubes, they take half Crystall Fritt, and half *Rochetta* Fritt, from whence notwithstanding proceeds a fair Sea-green, although in Crystall alone it ariseth most fair.

Skie Colour or Sea-Green.

CHAP. XXIII

SEt in the furnace a pot of pure metall of Fritt from *Rochetta*, or *Barillia* of *Spain*. The *Rochetta* of the *Levant* does best. As soon as the metall is well purified then take to a pot of 20 pound six ounces of Brass calcin'd of it's self, as in 20 *Chap*. put it into the metall as is said in the Skie colour in every particular; observing that this metall be skummed very diligently with a ladle. At the end of two hours the metall must be very well remixed; take thereof a proof, and leave it so 24 hours, then it is mixed, and wrought, and this will be most fair and marvellous Skie-colour varied with other colours, which are made in the art of glass. This colour tinged many pots in *Pisa* in the year 1602 and there came out a fair colour bearing all proofs.

A Red colour from Brass for many colours.

CHAP. XXIV.

TAke Brass in small plates, and put them on the arches of the furnace, leave them there closed until they are well Calcined of themselves with a simple fire, but not to melt. As soon as it is Calcin'd & powder'd it will become a red powder, for many, and those all necessary uses in the art of glass.

Brass thrice Calcined to colour glass.

CHAP. XXV.

PUt the said Brass into the *Fornello*, on the Lere of the furnace neer to the occhio, into earthen tiles or pans baked, Calcine it four days together, and it will become a black powder, and stick together, powder it again, serce it fine, and Recalcine it as before four or five days, for then it will not stick together, nor become so black, but russet, and powders of it self wherewith is made a Sea-green, and Emerald-green, the *Arabian* colour called *Turcois*, a very beautiful Skie colour, with many others. Wherefore observe that it be not too much nor too little Calcined at the third Calcination, for in this case it colours not the glass well, & the sign, that it is done perfectly is, that being put upon purified metal it makes it swell & boil suddainly and when it doth not so it is not good nor well Calcin'd.

A Sea-green in Artificial Crystal.

CHAP. XXVI.

TO a pot of Crystal Fritt containing 40 pound not charged with any *Manganese*, but well scummed, For when you would make a Sea-green, never cast the Crystal into water, but onely scum it carefully when this Fritt is melted and well purified, take 12 ounces of thrice Calcin'd brass and therewith mix half an ounce of *Zaffer* prepared, unite these two powders well together and put this mixture into the pot at four times, for so the glass receiveth it better. Mix the glass and powder with diligence, let them stand two hours, then remix them & take a proof, & if the colour be full enough let them stand; And although the Sea-green appears too full, yet the salt which is in the glass will eat up and consume the said greeness, and will always incline it to a blewishness. And at the beginning of 24 hours it may be wrought.

    I have many times experimented this way

of making Sea-green without ever erring. And if a moytie of *Rochetta* first be mixed with Crystal Frit, there will arise a fair Sea-green, and in Crystal alone 'tis marvellous fair.

General observations for all colours.

CHAP. XXVII.

THat the colours may arise in full beauty and perfection observe that every pot great or small, that is new, and put the first time into the furnace, leaves a foulness in glass from it's terrestriety, so that all the colours that are made in them appear bad and foul; wherefore those pots that are very great may be glased with white melted glass, as the Conciators well know, but the second time the pots lose this foulness. Observe secondly, that those pots which serve for one colour must not be used for another, for example a pot which hath been used for yellow, is not good to make a grain colour, and that which makes a grain-colour is not good to make a green-colour and that which serves for a red is not good to make a blew, and so of all other colours. Therefore every colour must have it's own pot, for in this manner colours will become more perfect. Thirdly, that the powders be well Calcin'd neither too much, nor too little; for in either of these cases they do not colour well. Fourthly, that a due proportion, and dose be used, and the mixture be made in proportion, and the furnaces be hot and fed with dry and hard wood. Fifthly, that the colour must be used dividedly, to wit, one part in the Fritt and the other in the metall, when it is melted and purified. There are other observations also which shall be treated of in their places, when we treat particularly of colours.

To make Copper thrice Calcin'd with more ease and less charge than the former.

CHAP. XXVIII.

TAke the Scales which the Brasiers make when they hammer pans, kettles, or other works of brass, which being often put into the fire the workmen hammer them, and these scales fall off, which cost much less than solid brass, wherewith is made the stuff hereafter described. And to Calcine it, you need not open and shut again the arches of the furnace (as in the aforementioned way) a thing of much disadvantage and disturbance to the furnace. Take then those scales that are clean, and free from all earth and foulness, wash them with warm water many times from their filth and uncleaness, and then let them be put into pots and pans of baked earth, and be kept in the leer near the Occhio, or in furnaces made for this purpose. I made at *Pisa* a little furnace in the fashion of a little

Calcar, where were calcin'd 20 or 25 pound of these Scales in few hours. Now let them stand in the said place four days, then renew them, powder and serce them fine, then again put them in the pans and pots of earth as before, with the same fire and heat as aforesaid for four days more, and they will come into a black powder, and run into lumps, beat, and serce those lumps fine, and repeat this process again, and a third time, then the Scales will be prepared with much less charge than the former and will thorowly have the same effects as the former; these scales rise much, wherefore use the prescribed care.

A fair Sea-green in Crystal metall, with the above-said scales.

CHAP. XXIX.

TAke a pot of sixty pound of Crystal Fritt well scummed, and not cast into water. I made a Sea-green without wetting the Crystall metall, and thought that it came out better. But we may make tryal of both ways, and stick to the best. Then take of metall well purified the said 60 pound, and one pound and a half of the said scales made with less charges, four ounces of *Zaffer* prepared, mix well these two powders together, put them to the Crystall at four times, mixing well the powder with the metall for two hours <u>then</u> put it again to be well remixed as 'tis usual, and take a proof. Herewith I have made many times a most fair Sea-green with happy success. Mix half Crystall with *Rochetta*, and you shall have a Sea-green every way beautiful.

A Sea-green of lesser charge.

CHAP. XXX.

TAke the same Brass prepared (as before) with the same quantity of *Zaffer*, put them in the same manner and form to the *Rochetta* of the *Levant*, and also to that of *Spain*, neither of which hath had any *Manganese*, and which hath been well skummed, and not passed thorow water using the rules as abovesaid in Crystal; for by this means it will receive a very fair Blew for all sorts of works, and will cost much less than Crystall: for the *Rochetta* is of much less value than the Crystall, as it is known. In this manner have I often made it at *Pisa*, and always with good success.

A marvellous Sea-green, above all Sea-greens, of my invention.

CHAP. XXXI.

Let the *caput mortuum* of the spirit of *Vitriol* of *Venus* Chymically made without corrosives stand in the air some few days; draw from it of it self without any artifice a green pale colour, this material being pulverised with the addition of *Zaffer* prepared, and with the same proportion (as is said in the other prepared Brass) the metall being added (as in the other Sea-green) it will make a Sea-green, so fair and marvellous that 'twill seem a very strange thing: I have often made it at

Antwerp to the wonder of all the spectators that saw it. The manner of making *Vitriol* of *Venus*, without corrosives, *Spagirically*, is to take little thin pieces Brass of the bigness of half a *Florentine* and to have one or more pots (as it is needful) and in the bottom of them to put a layer of common Brimstone powdr'd and above it little pieces of the Brass aforesaid, and <u>then</u> a layer of Brimstone and after that pieces of Brass, work in this manner till all the Brass that you have be set to work, this being done, let the Brass be baked as followeth in the 104 Chap. then prove it, and to your content you may see a thing of astonishment. I know not whether any have tried this way, which have found wonderful wherefore I say 'tis my own invention.

A green Emerald colour in glass.

CHAP. XXXII.

IN making Green you must observe that the metall have not much salt, with metall that hath much salt as Crystall and *Rochetta* have, you cannot make a fair Green, but onely a Seagreen, for the salt consumes the Green and always inclines the colour to a Blew. Wherefore when you would make a fair Green put common metall made with *Polverine*, into small or great pots, and in no wise have any *Manganese*. When it is melted and well purified, add to this metall a little *Crocus Martis* calcin'd with vineger, about three ounces thereof to a hundred weight, let the metall be well mixed, and remain so an hour until the metall incorporate the same tincture of the Crocus, which will make the glass come out Yellowish, and takes away the foulness and Blewness, which the metall always hath. This process, will give the metall a fair Green. Put of thrice calcin'd

Brass, made with scales (as before) two pound to every hundred pound of metall, and this must be added at six times mixing well the powder, with the metall, then let them settle two hours, and the metall incorporate with it, then mix again the metall, and take a proof, and if the Green enclines to a Blew add a little more *Crocus Martis*, so you shall have a very fair Sea-green, called Leek green, which at the end of twenty four hours may be wrought: This Green I have many times made at *Pisa*, which came forth sufficiently fair. And so it will to every one that shall observe punctually what is abovesaid.

A Green fairer than the former.

CHAP. XXXIII.

But if you would have a Green much fairer and shining than the former, put into a pot of Crystalline which hath not had any *Manganese*, and which hath passed thorow water once or twice, till all the saltness be gotten out, and to the

Crystalline, let half of common white metall made of *Polverine* be put in at several times, as soon as this metall is well mixed and purified, take to every hundred pound, two pound and a half of thrice Calcin'd brass, made with plates of Brass in the arches of the furnace, and with this mix two ounces of *Crocus Martis* calcin'd with Brimstone, and reverberated, put these two powders well mixed together to the abovesaid metall, using the rules as before in the said Green, if the metall hath any Blewness give it a little of the said *Crocus Martis*, which takes it away, and then work it as the other Greens, and there shall be made the wonderfull Green of the *Burnet*. I have thus made it many times at *Pisa* with very good success, for works more exact than ordinary. If you will have a fair colour, see that the Brass be well prepared.

A marvellous Green.

CHAP. XXXIV.

TAke Brass thrice calcin'd as before then in stead of *Crocus Martis*, take the scales of iron which fall from the Smiths anvils, powder them finely, sift them clean from the coals and ashes, and with the quantity aforesaid, mix them well with the Brass, and put them to the common glass metall of *Polverine*, without any *Manganese*, with the rules aforesaid in the Green, and with this *Crocus Martis*, or scales you shall doubtless have a more marvellous Emerald Green-colour, which will have wholly lost it's Azure and Sea-colour, and will be a Yellowish green after the Emerald, and will have a shining and fair lustre than the aforesaid Green. The putting in of scales of iron was my own invention. In the rest of the work let the rules and doses as in other Greens be observed and you shall have a strange thing, as experience hath often shown me.

Another Green, which carries the Palm from all other Greens, made by me.

CHAP. XXXV.

TO a pot of 10 pound of metall to wit half of Crystalline passed thorow water several times, and half of common white metall of *Polverine*, take four pound of the common Frit of *Polverine*, wherewith mix three pound of red Lead, unite them well together, and put them into the same pot, and in few hours all of them will be well purified, then cast all this metall into water, and take out the Lead, then return the metall which hath passed thorow the water into the pot, & let the metall purifie for a day, then if you put in the colour made Chymically with the powder of the *Caput mortuum* of the Spirit of *Vitriolum Veneris*, adding a very little *Crocus Martis*, there will arise a marvellous Green, fairer than ever I made any, which will seem to be a very Emerald of the ancient Oriental rock.

A Blew or Turcois, *a principal colour in this art.*

CHAP. XXXVI.

PUt sea salt which is called black or gross salt (for the ordinary white salt which is made at *Volterra* is not good) into the Calcar or Fornello till all the moisture be evaporated, and it becomes white, then pound it well, to a small white powder. The salt so calcin'd, keep to make a Blew or *Turcois* colour. Put into a small or great pot Crystal metall died with the colour of Sea-green (made, as hath been said many ways.) But let the colour be fair and full (for this is of great importance to make a fair Skie colour) according as you would have the Sea-green fair and excellent. To this metall so coloured, put of the said salt calcin'd into the pots, mixing it well with the metall, and this is to be put in by little and little until the Sea-green lose its

The first Book. 57

transparencie, and diaphanietie, and takes opacity, for the salt being vitrified makes the metall lose it's transparencie, and gives it a little paleness, and so by little and little makes the said Skie colour, which is the colour of a *Turcois-stone*; when the colour is enough it must be wrought speedily, for the salt will be lost and evaporated, and the metall returns again to be transparent and foul-coloured. But when the colour is lost in working add new burnt salt (as before) that the colour may be reduced, and so you shall have your desired colour. Let the Conciators well observe that this salt always crackles when it is not well calcined, therefore let him have a care of his eyes and sight, for it endangers them. The quantity of salt must be put in by little and little, leaving some distance between each time, till he see the desired colour. But in this I used neither dose nor weight, but my eye onely. I have often made this colour, for it is very necessary in beads,* and the most prised and esteemed colour that is in the art. Wherefore to make a Blew for beads, take

---

* Correction of a rare mistranslation by Merrett

the Green of Crystal metall, and half Sea-green made of half *Rochetta*, which will become a fair colour, although it be not all Crystall metall.

The second Book, wherein are shown the true ways of making Calcidony *of the colour of* Agats, *& oriental* Jaspers, *with the way to prepare all colours for this purpose, and also to make* Aqua-fortis, *and* Aqua Regis *necessary in this business.*

And the Manner of calcining Tartar, *and uniting it with* Rosichiero, *made* Chap. 128. *Which produceth pleasant toyes of many colours with undulations in them, and gives it an opacity such as the Natural and Oriental stones have.*

CHAP. XXXVII.

SInce I am to shew the manner how to make *Calcidonies, Jaspers* and Oriental *Agats* it is necessary first to teach the preparation of some mineral things, for such compositions, and although some them may be publiquely bought, yet notwithstanding, I being desirous that the work should be perfect judged it pertinent to my purpose to shew the most exquisite Chymical way, that the skilful may make every thing of themselves, both more perfect. and with lesser charge. For there is no doubt that when the materials are well prepared, and the colour of the metalls is well opened, and separated from their impurity and terrestriety which usually hinder the ingress of their tincture into glass and their union in their smallest parts, that then they colour the glass with lively, shining and fair colours, which very far surpass those that are vulgarly, and usually made in the furnace. And because the colour of *Calcidony*, or rather it's compound (which is nothing else, but as it were a reuniting of all the colours, and toyes that may be made in glass, a thing not common nor known to all) if they not well prepared, and subtilised as is necessary, they give not the beauty and splendor to glass as is required. Wherefore it is necessary that the metalls be well calcined subtilised, and opened with the best *Aqua-fortis, Sulphurs, Vitriols, sal Armoniak*, and suchlike materials, which in length of time, and at a gentle heat, are opened and well prepared, but a violent fire herein hurteth much. *Tartar* and *Rosichiero* (besides their being very perfect and well calcined) must be also put in proportion and in fit and due time, and you must also observe, that the metall be well boiled, purified, and perfected, and in working of it some such care is to be used, as the diligent masters are wont to use, and by thus the true *Jasper* and *Agat*, and Oriental *Calcidonies*, with the fairest and beautifullest spots of wavings, and toyes, with divers lively and bright colours. Hence it truly appears that nature cannot arrive so high in great pieces, and although it is said and may be made to appear true, that Art cannot attain to Nature, yet experience in many things shews, and in particular in this art of the colours in glass, that art doth not onely attain to and equal nature, but very fair surpasses and excells it. If this not were not seen, hardly would you believe the beauty the toyes and wavings of divers colours, variously disjoyned one from the other with a pleasing distinction, which is seen in this particular of the *Calcidony*. When the medicine is well prepared, and the glass wrought at a due time, the effect that cometh thence passeth all imagination and conceit of man. In the three ways to make it, which I teach, I believe you may see how far the art of glass ariseth in this particular, where I demonstrate every particular so distinctly, that any practitioner, and skilful person, may understand and work without errour, and he that works well may find out more than I set down.

How to make Aqua-fortis *call'd parting water, which dissolveth silver and quick-silver, with a secret way.*

CHAP. XXXVIII.

TAke of Salt-peter refined one part, of Roch-alum three parts; but first exhale in pans all the humidity from it; to every pound of this stuff add an ounce of Crystalline Arsnick (this is a secret and no ordinary thing) which besides it's giving more strength to the water, helps to extract better the spirits from the materials, which are the true nerves and strength of the *Aqua-fortis* without which the water perhaps would be no better than well-water. Powder and mix them well together adding thereunto the tenth part in the whole of Lime, well powdred, mix them well, and put so much of this stuff into glass bodies, that about three quarters of them may be full, let them be luted with strong lute, which I remit to the Artist as a common thing: but one not vulgar I will declare. Take some lome for example of the river *Arnus* (which is a fat earth known to all) one part, of sand 3 parts, of common wood-ashes well sifted, of shearings of woollen cloath, of each one half; mix them well together, and incorporate them into a past with common Water, work them well together, for the more 'tis wrought the better 'tis, therefore see that your past be a little hard, to all these add a third of common salt, which incorporate well with the lute, 'tis a business of importance, then lute the glasses with this perfect lute, and set them in wind furnaces, fitting to their bottoms, baked earth which will bear the fire. Under the bottom of these bodies, let there be four fingers of sand, & thick Iron bars to bear the weight, & fill'd round about with sand, put receivers of glass to them, large and capacious within, lute the joynts well with lute made of fine flowre and lime, of each a like quantity powdred, mixed, tempered, and impasted with the whites of Eggs well beaten, with this lute, binde and lute the joynts with roulers of fine linnen, which, when well dryed and rould about three or four times make a very strong lute, rouling but once at a time, and letting it dry a little before the Second rouling. And then this will bear all the violence fury, and force of the spirits of the *Aqua-fortis*, and to this end fit exactly a very large receiver to every glass body. And when they are well dryed make a fire in the furnace onely with coal at first, and that a very temperate one, for three hours, for in that time the windy moisture distilleth off, which endangers the breaking of the glasses, and continue for six hours a moderate fire, afterwards encrease it gently, adding billets of dry oaken wood to the coals, and so proceed by little and little, augmenting it for six hours more, and then the head will be tinged with Yellow, a sign that the Spirits begin to rise; continue this degree of fire untill the spirits beginning to condensate colour Red the receiver and head, which will always grow deeper colour'd like a Rubie. Then encrease the fire for many hours, till the head and receiver become Red, which sometimes lasteth two whole days. Continue the fire by all means till all the Spirits of *Aqua-fortis* be distill'd off; which is known, when the head & receivers by and little, begin to grow clear, and become white as at first, and wholly cold; yet notwithstanding continue the fire one hour more. Then let the furnace cool of it's self. Observe that when the head and receivers are Red, and the fire strong, you admit no wind, nor cold air into them, nor touch them with any cold thing, for then they will easily crack, and your pains, cost and time will be lost, wherefore when they are in this state, let them be kept hot in the fire. Now, when all is cold, put upon the head and receiver linnen cloaths wetted and well soaked in cold water, that the spirits which are about the head and receiver may the better sink into to the *Aqua-fortis*, leave them thus for 12 hours, then bath the joynts and lutings with warm water, till they being moistned you may take off the bandage, and the head from the receiver, which usually are safe. The bodies may be broke and thrown away, for they will serve no more, powder the dregs and residences of the *Aqua-fortis*, to wit about their third part, and to every pound of them add four ounces of Salt-peter refined, and put them into another body luted, and pour on them the said *Aqua-fortis* lute and distil them as before in everything. Keep the *Aqua-fortis* in earthen jugs well stopt that the better spirits may not evaporate. This parting water is good for the following uses. Some there are that instead of *Roch Alume* take as much more of the best *Vitriol*, such as the *Roman* or the like is. The sign that Vitriol is good for this use, is, that being rub'd upon polished Iron it colours it with a Copper colour. Vitriol purified after the following manner will make a stronger *Aqua-fortis* than Alume.

To purifie Vitriol *to make the strongest* Aqua-fortis.

CHAP. XXXIX.

Dissolve the best Vitriol (the better, the stronger the Aqua-fortis) in common water, let the solution stand three days being impregnated with salt, then filtre and evaporate in glass bodies two thirds of the water, and put the remainder into earthen pans glased, which set in a cold place, in 12 hours the *Vitriol* will shoot into pointed pieces, appearing like natural Crystall of a fair Emerald colour. Dissolve this same Vitriol again and do as before, repeat it thrice, at each solution there will remain at the bottom of the glass a Yellow substance, which is it's unprofitable Sulphur, and is to be cast away. At the third time the Vitriol will be purified, and fit to make a good and strong *Aqua-fortis*, much stronger than the ordinary, especially if the Nitre be well refined.

How to make Aqua Regis.

CHAP. XL.

TO every pound of the said *Aqua-fortis* put two ounces of sal Armoniack powdered, into a glass body, which set in a pan full of warm water, and let the *Aqua-fortis* be often stirred, which will soon dissolve the *sal Armoniack* with it's heat, which will be tinged with a Yellow colour, put in more *sal Armoniack*, as long as the Aqua-fortis will dissolve any, when it dissolves no more let it settle a little, when it is clear decant it leasurely off, and in the bottom there remains the unprofitable terrestriety of the *sal Armoniack*. Now this Aqua Regis is strong and fit to dissolve Gold, and other metalls but silver it toucheth not at all.

To burn Tartar.

CHAP. XLI.

PUt *Tartar* of Red-wine which is in great pieces, and appears full of spots (lay by that which is in powder, for it is not good) into new earthen pots, and let it burn in kindled coals until it smoaks no more, and when it is calcin'd, and in lumps of a black purplish substance then it is burned and prepared.

How to make a Calcidony *in Glass very fair.*

CHAP. XLII.

Ut of *Aqua-fortis*, two pound, into a glass body not very great, but with a long neck, four ounces of fine silver, in small and thin pieces, and set them neer the fire, or in warm water, which as soon as the *Aqua-fortis* begins to be hot 'twill work and dissolve the silver very quickly and continue so until it hath dissolved, and taken it up, then take a pound and a half of *Aqua-fortis*, and in it dissolve (as you have before done with silver) six ounces of Quick-silver, when all is dissolved let the two waters be well mixed in a greater body, and powr upon them six ounces of *sal Armoniack*, and dissolve it at a gentle heat, when it is dissolved put into the glass one ounce of *Zaffer*, and half an ounce of Manganese, each prepared, and half an ounce of *Ferretto* of *Spain*, a quarter of an ounce of *Crocus Martis* calcin'd with Brimstone,

The second Book. 71

thrice calcin'd Copper, Blew smalts of the Painters and Red lead, of each half an ounce, powder all these well, and put one after another into the body, which then stir gently that the *Aqua-fortis* may be incorporated well with the said powder, keep the body close stoped for ten days, every day stirring it well several times, and when they are well opened, then put it into a furnace on Sand, and make a most temperate heat, so that in 24 hours all the *Aqua-fortis* may be evaporated. Observe that at last you give not a strong but a gentle heat, that the Spirits of the *Aqua-fortis* may not evaporate, and in the bottom there will remain a Lion colour, which being well powdered, keep in a glass vessel. When you would make a *Calcidony*, put into a pot very clear metall and made of broken pieces of Crystall vessels, and Crystalline, and white glass which hath been used. For with the Virgin *Fritt* which hath never been wrought, the *Calcidony* can never be made, and the colours stick not to it, but are consumed by the *Fritt*. To every pot of about 20 pound of glass, put two ounces, or two ounces and a half, or three ounces of this powder, or medicine, at three time and incorporate, and mix them, that the glass may take in the powder, in doing whereof it raiseth certain Blew fumes, as soon as it is well mixed let the glass stand an hour, then put in another mixture, and so let it alone 24 hours, then let the glass be well mixed, and take thereof an essay which will have a Yellowish Azure colour, this proof being returned many times into the furnace, and taken when it begins to grow cold, will shew some waves, and divers colours very fair. Then take *Tartar* eight ounces, soot of the Chimny well vitrified two ounces, *Crocus Martis* calcin'd with Brimstone half an ounce, put by little and little all these well powdered and mixed into this glass at six times, expecting a little while at each time, still mixing the glass that the powder may be well incorporated. As soon as all the powder is put in, let the glass boil and settle 24 hours at least, then make little glass body of it, which put in the furnace many times, and see if the glass be enough and if there be on the outside toyes of Blew, and Seagreen, Red, Yellow, and all colours with toys, and it hath some waves, such as *Calcidony, Jaspers*; Oriental *Agats* have, and that the body kept within be as to the sight as red as fire. Now as soon as it is made and perfected, it is wrought into vessels always variegated, which are not to be remade, for they do not arise well. These vessels may be made of divers sorts, and drinking glasses of many fashions, broad drinking cups, salts, flower pots, and the like toys, still observing that the master workman pinch off well the glass (that is wrought) with pincers, and anneal it sufficiently, that it may make waves and toys of the fairest colours. You may also make with this pot dishes, pretty large in *Oval, triangular, quadrangular* form, as you will, and afterwards work it at the wheel (as Jewels) for it takes polishing, and a fine lustre, and it may serve for little tables, and cabinets; so that those little Jewels will represent the Oriental *Agat, Jasper*, and Oriental *Calcidony*, and when it happens that the colour fadeth, and the glass becomes transparent, and no more Opacous as it ought to be for these works, then cease from working, put to it new *Tartar* calcin'd, soot

74 *The second Book.*

and *Crocus*, for thus (as before) it takes a body and Opacousness, and makes the colours appear; set it then to purifie many hours, that the new powder put in may be <u>incorporated</u>, as 'tis usual, then work it. This was my way to make Calcidony in the year <u>1601</u> in *Florence*, at *Casino* in the glass furnaces; at which time I caused to work in the furnace, the brave Gentleman *Nicolao Landiamo* my familiar friend, and a man rare in working Enamels at the lamp, in which furnace I made many cups of *Calcidony* at the same time, which always were fair to all essays, never departing from the aforesaid rules, and having the materials well prepared.

The second Calcidony.

CHAP. XLIII.

1. IN a pound of *Aqua-fortis* dissolve three ounces of Silver cut small in a glass body well closed, set this aside.
2. In another glass body, put one pound of *Aqua-fortis*, wherein dissolve 5 ounces of *Mercurie* well purified, close the body well and set that aside.
3. In another little glass body, put one pound of *Aqua-fortis*, wherein dissolve two ounces of *sal Armoniack*, then put into this dissolution of *Crocus Martis* made with *Aqua-fortis*, *Ferretto* of *Spain*, Copper calcin'd Red, as in Chap. 24. Brass calcin'd with Sulphur, of each half an ounce, put all these materials well ground, and powdered by them-selves, and then one by one, into the bodies by little and little, with patience, because they all arise much.
4. In another little glass vessel let there

be put one pound of Aqua-fortis, and therein dissolve one ounce *sal Armoniack*, and in the dissolution of crude *Antimony* powdered, *Vitriol* purified, of Azure, or Blew Smalts, of each half an ounce, one ounce of Red-lead, grind them all well and set the vessel by.

5. In a like body, dissolve in one pound of *Aqua-fortis*, two ounces of *sal Armoniack*, then put in one ounce of *Zaffer* prepared, and a quarter of an ounce of *Manganese* prepared, and half an ounce of thrice calcin'd Copper, and one ounce of *Cinaber*, put in warily every one of these things (well powdered) by themselves, into the body, avoiding those things that swell up arise and fume, set this aside.

6. In another small glass body, dissolve in one pound of *Aqua-fortis* two ounces of *sal Armoniack*, and then add of *Cerus*, Painters Red-lake, *Verdigreas*, the Skales of iron that fall from the anvil, of each half an ounce, these swell very much. Let all these 6 bodies stand 12 days, & shake them well six times every day, that the water may penetrate & subtilise the ingredients and metalls, to communicate their tincture to the glass.

The second Book. 77

After this time take a great glass body, luted at the bottom, into which you shall empty all the materials of these six bodies by little and little, that they may not run out, nor make the glass crack, in this great body mix well the waters, that the materials may be well united, and mixed together, set this glass in ashes at a very gentle heat, for twenty four hours, that the water may evaporate. Observing that the fire be gentlest at last, lest the powder be wasted with too much heat. He that will regain the *Aqua-fortis* may joyn the head & receiver & lute the joynts (as is usual) and the water being evaporated, there remains at the bottome a reddish powder, which is kept in a glass closed for use.

Put this powder or medicine into metall made of broken pieces of glass, and old glass, but not made of Virgin *Fritt* of Crystall, or Crystalline, as in the first *Calcidony* hath been said. Give the metall the same quantity, and use the said distance of time as in the other, then give it the body of burnt *Tartar*, and soot of the Chimny *Vitrified*, and *Crocus Martis* made with vinegar, then let them settle.

twenty four hours, and cause a vessel to be made thereof, and put it in the fire and observe whether it take body and opacity, and if it shew the variety of colours with toyes and wavings, work all of it into vessels of divers sorts, framing therewith all sorts of workmanship and variety of toyes.

With this sort of Calcidony, I made many cups, all which were fair, and besides with this past of Calcidony may be made many hundred crowns for gentle men, as fair as can be uttered. These were seen by *Ferdinando* the great Duke, of blessed memory, and by many other Princes, and Lords, and this was done by me in *Flanders*.

The second Book. 79

The third way of Calcidony.

CHAP. XLIV.

1. IN a glass body in strong *Aqua-fortis*, dissolve four ounces of fine leaf Silver, that is to say, round cuttings of leaf silver, stop the body and set it aside.

    2. In another body of like glass dissolve in one pound of *Aqua-fortis* five ounces of *Quick-silver* purified with vinegar and common salt, in a wooden dish with a wooden pestle stir the *Mercury* sufficiently round with strong vinegar, and wash it with clear common water, until 'tis dissolved, and carry away all the common salt, together with the blackness of the *Mercurie*, repeat this many times. Then strain this *Mercurie* through canvas, and dissolve it in the above said *Aqua-fortis*, as before, close the glass vessel and set it aside.

    3. In another glass body, dissolve in a pound of Aqua-fortis, three ounces of fine

80　　*The second Book.*

Silver calcin'd after this manner, to wit, amalgamate the silver with *Mercurie*, mix the amalgama with as much more common salt well prepared from all 'its terrestriety, by dissolving it in common water, and boyling it a little, and then let it settle two dayes that the terrestriety mixed with the salt may sink to the bottome, then filtre the water, and in the bottome will remain the grossness and terrestriety of the salt, evaporate this water filtred from the terrestriety of the salt in a glass vessel, and dry it well, repeat this till the salt sends no more dregs to the bottom, and then it will be perfect and fit for the work. This purifying of the salt is made that it may be more efficacious to open the silver, otherwise it will be hard to separate them. Put all these things amidst the coals, in a pot, that all the *Mercurie* may be evaporated away, and the Silver remain at the bottome calcin'd and powdered, and add unto it it's weight of new common salt prepared (as before) mix them well, and put all in a chrysible or a pot to calcine six hours in the fire. Wash this stuff in a glased pot many times with warm water till all the saltness be well gone, then put this silver into a glass body full of common water, boil it till a quarter of it be evaporated, then let the silver grow cold and settle, and decant the water, repeat this fresh water thrice, and the fourth time put it in a body of *Aqua-fortis*, stir it well, and set it aside.

4. In another like body, dissolve in a pound of *Aqua-fortis*, three ounces of *sal Armoniack*, decant off the clear solution, the remainder at the bottom cast away. In this water dissolve a quarter of an ounce of gold, keep this last solution apart.

5. In another glass body, dissolve in one pound of *Aqua-fortis*, three ounces of *sal Armoniack*. Then put into the solution, of *Cinaber*, of *Crocus Martis*, of *ultramarine*, *Ferretto* of *Spain*, of each half an ounce, put them (well powdered) leasurely into body, which being done close the vessel, and set it aside.

6. In another body, dissolve in a pound of *Aqua-fortis*, three ounces of *sal Armoniack*. Then put in *Crocus Martis* calcin'd with vinegar, calcined Tin, a thing common in potters furnaces, *Zaffer* prepared, and *Cinaber*, of each half an ounce. Put gentlie each of them (ground by themselves)

82 *The second Book.*

into the *Aqua-fortis*, then keep this in a vessel, and set it aside.

  7. In another body of glass dissolve in a pound of *Aqua-fortis* two ounces of *sal Armoniack*. Then put leasurely into the solution, Brass calcin'd with Brimstone. Brass thrice calcin'd, as in Chap. 28. *Manganese* prepared, and the scales of Iron which fall from the Smiths anvil, of each half an ounce. Put each of these well ground by themselves, by little and little then close the vessel, and set it aside.

  8. In another body, dissolve in a pound of *Aqua-fortis*, two ounces of *sal Armoniack*, whereto put *Verdigreas* one ounce, Red-lead, crude Antimony, and the *caput mortuum* of *Vitriol* purified, of each half ounce, put these powdered leasurely in, close the vessel, and set it aside.

  9. In another body, dissolve in a pound of *Aqua-fortis*, two ounces of *sal Armoniack*, then put in leasurely *Orpiment*, white *Arsnick*, Painters Lake, of each half an ounce, each powdered, and ground by its self, close the vessel, and set it aside.

  Keep these nine bodies (well closed) in the furnace fifteen days, and every day stir it well many times, that the *Aqua-fortis* may work, and the materials be subtilised, and their tinctures well opened, else they will not work well, then put all the materials with their waters into a great and strong body, by little and little; the things being united together, let alone the great body (whereinto you have powred the materials of all the lesser bodies) closed for six dayes, and every day stir it, then put it in ashes, giving it a gentle heat for twenty four hours, that the water may onely evaporate, observing that the body must be well luted at the bottome, even unto the midst of the body, and at the last of all the heat must be made so gentle that it onely evaporate the water and that the better spirits of the *Aqua-fortis* may remain inclosed in the same powders, for so the powder will work fair, and strange things in glass. In the bottome of this body, will remain a powder of a purplish Green, whereof I gave the glass such a dose and quantity as is said in the first *Calcidony*. Then in due times (as is said in the first *Calcidony*), give it it's body, to wit, *Tartar* burnt, the soot of the Chimny, and *Crocus Martis* made with vinegar, using the same dose, and diligence, and times, and intervals throughout, as is said in the first *Calcidony*, then at the end of twenty four hours, work it with diligence, and according to art, and set it to the fire again, as hath been most punctually said in the first *Calcidony*.

    This third way of making *Calcidony*, I performed at *Antwerp*, a City of *Brabant*, *Anno* 1609. in the Moneth of *January*. At which time, and for many years, there was in the house Signor *Emanuel Nimenes* a Knight of the Noble Religion of Saint *Steven*, a *Portughes*, and Citizen of *Antwerp*, a gentile Spirit, and Universal in all knowledge, as any in the Low-Countries, whom I saw or knew. With this powder I made a *Calcidony* in the furnace of Antwerp, which I caused Signor *Philippo Ghiridolpho* a very Courteous Gentleman to work, which *Calcidony* came forth so fair, and beautiful, that it imitated the true Oriental *Agat*, and in fairness and beauty of colours far surpassed it. Many *Portughes* Gentlemen well Skilled in Jewels admired it, saying that nature could not do more. This was the fairest *Calcidony* that ever I made in my life, which although

The second Book.

it be laborious, and long a working, yet notwithstanding it doth real things. Of this *Calcidony* two vessels were given to the most Excellent Prince of *Orange*, which pleased him very well.

The third Book. This Book shews the wayes to make the colour of Gold Yellow of the Amethist, Saphyre, Granat, Velvet Black, Milk White, Marble and Deep Red; As also to make Fritt *with natural Crystal, and to colour glass of a Pearl colour, and other particulars necessary in this Art.*

CHAP. XLV.

THis third Book teacheth various wayes, and one better than another, to make all the abovesaid colours; As also a particular way to make Fritt of natural Crystal, which will melt as ordinary Crystal metall, and will make vessels very white, beautiful, and sightly. There is no doubt but some of those colours are known to Artists, though not to all persons. For few they are that know how to make well Gold Yellow, and a Deep Red, being hard and nice colours in this Art. Since in making them 'tis necessary you be punctual in the dose, time, circumstances, and materials: for if you err but a very little in any of them whatsoever, all the whole labour and business is lost and comes to nothing. I describe these two colours, and all other, in so clear and intelligible a stile that every body may understand, and make them to their gust and satisfaction.

1. *You must be exact in the time, quantity, circumstances, purifying, powdering, sercing, fire, materials, if you err but a little in any of them whatsoever, all the labour is lost, and the colours come to nothing.*

2. Tartar *must be of Red-wine well vitrified in the vessel, in gross pieces, not in powder Vitrified naturally of themselves. That of white wine is not good.*

3. *To* Manganese *our author still subjoyns of* Piemont.

4. *The colour must be made fuller or lighter according to the works you employ them for, and to heighten them, put in more of the colour, but to make them lighter put into the pot more* Fritt. *Take some metall out of the pot, and you shall see whether you have your desired colour; put in your colours by little and little lest they overdo.*

5. *Put your colour to the* Fritt, *and not to the metall, when melted, for then it neither takes the colour so well, nor so good a colour.*

6. *Mix the colours well with the metall in the pots, when 'tis melted, that both may be well incor-porated, and this is to be done often as you work the metall.*

To make a Gold Yellow in glass.

C H A P. XLVI.

TAke Crystal *Fritt* two parts, *Rochetta Fritt* one part both made with *Tarso*, (which is much better than sand) mix and remix well these two *Frits*, and to every hundred pound of this composition, take of Tartar in lumps well beaten and serced fine, of *Manganese* prepared, of each one pound, mix these two powders well, first together, and then with the Frits. Then put them into the furnace, and let them stand four days at an ordinary fire, because they rise much. When the metall is purified and well coloured (which usually is the end of four days) work it into vessels and works. This quantity of the materials makes a most fair colour, which you may make deeper or lighter by adding or diminishing the powders or Frits. You must put the powder in at several times, and not into the metall, for then it colours not. With these rules amid observations you shall make a very fair Gold Yellow. But if you would have it fairer and a more graceful Yellow, take all Crystall *Fritt*, And thus I have frequently made this colour and alwayes very fair.

Garnat colour.

CHAP. XLVII.

TAke of Crystall and *Rochetta Fritt*, of each a like quantity, mix them well, to every 100 weight, add of *Manganese* one pound, *Zaffer* prepared an ounce, mix well these two powders together first, then with the *Frits*, then put this powder into the pot by little and little. Mix well the *Manganese* with the *Zaffer*, for this quickens the colour, making it shining, beautiful and fair. At the end of 24 hours (when 'tis pure and well coloured) work it.

Amethist colour.

CHAP. XLVIII.

TAke onely Crystal Fritt made with the most perfect *Tarso, Manganese* prepared one pound, *Zaffer* prepared one ounce and a half, mix these two powders well together, and then with the Fritt, and not with the metall in the pots. The proportion is one ounce of the mixed powder to one pound of the *Fritt.* When the metall is pure and well coloured work it into vessels, &c.

Saphyre colour.

CHAP. XLIX.

TO every hundred weight of *Rochetta Fritt*, add one pound of *Zaffer* prepared, & to every pound of *Zaffer* one ounce of *Manganese*, mix these two well together first, and then with the *Fritt*, put them all mixed into the furnace to melt and purifie, and when 'tis pure, and well coloured work it, &c. This small quantity of *Manganese* makes a most fair colour of a double violet, which I have often made at *Pisa* and always well.

A fairer Saphyre Colour.

CHAP. L.

INstead of *Rochetta Fritt*, take Crystal *Fritt*, whereto add the same quantity of the foresaid powder, with the same rules and you shall have a fair, and shining *Saphyre* colour.

A Black colour.

CHAP. LI.

TAke pieces of broken glasses of many colours, grind them small, and put to them *Manganese* & *Zaffer*, to wit, not more than half of *Manganese* to the *Zaffer*. This glass purified will be of a most fair Black, shining like velvet, and will serve for tubes and all kinds of works.

A much fairer Black.

CHAP. LII.

TAke of the *Frits* of *Crystal* and *Polverine*, of each 20 pound, Calx of Lead, and Tin four pound, mix all together, set them in a pot in the furnace well heated, and when the metall is pure, take steel well calcined and powdered, scales of Iron which fall from the Smiths anvil, of each a like quantity, powder and mix them well, put six ounces of this powder to the said metall that they may both strongly boil, let them settle 12 hours, and sometimes mix the metall, and then work it. This will be a most fair Velvet Black, and pleasant, to make all sorts of works.

Another fairer Black.

CHAP. LIII.

TO a hundred weight of *Rochetta Fritt*, give two pound of *Tartar*, and of *Manganese* six pound, both pulverised, mix them and put them in the furnace leasurely, let the metall purifie, which will be about the end of four dayes, then mix, and wash the said metall, will make a more marvellous black than all the former.

A fair milk White called Lattimo.

CHAP. LIV.

TAke of Crystal Fritt twelve pound, of calcined Lead and Tin two pound, mix them well, of *Manganese* prepared half an ounce, unite them all together, and put them into a pot heated, let them stand twelve hours that the materials may be melted, and at the end of eight hours you may work it. This will be a fair White which I have often made.

A fair White much whiter than the former.

CHAP. LV.

TAke 400 weight of Crystal *Fritt*, and 60 pound of calcined Tin, and two pound and a half of *Manganese* prepared, powder and mix them all with the *Fritt*, and set them in a furnace in a pot, let them refine, and at the end of 18 hours this stuff will be purified, which cast into water, purifie it again in the furnace, and make a proof, and if it be too clear add 15 pound of the aforesaid calcined Tin, mix well the metall many times, and at the end of one day it becomes marvellous white, and in whiteness surpasseth any snow, then work it. I have often made it and always with good success. This white may be also made with *Rochetta*, but not so white as with Crystal.

To make a Marble colour.

CHAP. LVI.

PUt Crystal Fritt in a pot, and when 'tis melted (before 'tis purified) work it. This is a fair Marble colour.

A Peach colour in White.

CHAP. LVII.

MAnganese prepared will make in Lattimo the colour of a Peach-flower. But work it in time because it loseth colour.

A Deep Red.

CHAP. LVIII.

TAke of Crystal *Fritt* 20 pound, broken pieces of white glass one pound, calcined Tin two pound, mix these well together, put them into a pot to run and purifie, when these are melted, take steel calcined, scales of Iron from the anvil, both well ground, of each a like quantity, mix them together, put leasurely of this mixture, about an ounce, to the aforesaid metall when purified, and mix them well, and let them incorporate, which succeeds commonly in five or six hours. Too much powder makes the metall black, whereas the colour ought to be transparent and not opacous, of an obscure Yellow; when 'tis so, put in no more powder, but then put about three quarters of an ounce of Brass calcined to redness (as in the 24 Chap.) and ground,

The third Book. **101**

to this metall, and mix them many times, and at about three or four times it will become as red as blood, wherefore make essays often, and see whether this colour be good, and when so, work it speedily, else 'twill lose it's colour, and become black. Besides leave the mouth of the pot open, else the colour will be lost. Let not the pot stand above 10 hours in the furnace, and suffer it not to cool as much as is possible. When you see the colour fade (which sometimes happens) put in some scales of Iron, which reduceth the colours. And, because this is a nice colour, use all diligence in making it by putting in the steel and scales, as also in working it.

Fritt *of natural Crystal.*

CHAP. LIX.

*C*Alcine natural Crystal in a Chrysible extinguish it in common cold water eight times, cover the Chrysible that no ashes nor filth get in, Dry the calcined Crystal, and grind it to an impalpable powder, mix this powder with salt of *Polverine* made in a glass body, as in Chap. 3. with these make a *Fritt*, observing the quantities, rules, and portion of *Manganese*, setting it in the furnace, & at due, and often times casting it into the water, purifying and working it as in other Crystal. And thus you will make a marvellous thing.

A Pearl colour in Crystal.

CHAP. LX.

PUT at 3 or 4 times to Crystal melted and purifed, of *Tartar* well calcin'd to whiteness, and continue to put in the *Tartar* 4 or 6 times, always mixing it well with the metall, till the Crystal hath gotten a Pearl colour. Then work it speedily, for this colour fadeth. This I have often practised and experimented.

[page 104 is blank]

The fourth Book. Wherein is shown the true way to make glass of Lead, to calcine Lead, and extract from it the colours of green Emerald, Topaz, Skie Colour or Sea green, Granat colour, Saphyre, Gold Yellow, and of Lapis lazuli.

With the way to colour natural Crystal (without melting it) into the permanent colours of Rubies, Balas, Topaz, Opal, Girasole, & other fair colours.

CHAP. LXI.

THe glass of Lead known to few in this Art, as to colours, is the fairest and noblest glass of all others at this day made in the furnace. For in this glass the colours imitate the true Oriental gems, which cannot be done in Crystal, nor any other glass. 'Tis very true, that unless very great diligence be used, all sorts of pots will be broken, and the metall will run into the coals of the furnace. Observe my rules in all these glasses made of Lead exactly, and you shall avoid all danger. This business principally consists in knowing well how to calcine Lead, and to recalcine it also a second time; For by how much 'tis better and more calcined, by so much the less it returns to <u>lead again</u>, and by consequence the less breaks out the bottom of the pot. Secondly, cast the metall into water, and separate carefully the Lead from the glass, even the least grains of it. This glass of lead must be cast into the water by little and little to make a better separation for the least Lead remaining breaks out the bottom of the pots, and lets all the metall run into the fire.

These two rules our Author repeats almost in every Chapter of this Book, and these following also,

The pots and Lead must not have too much heat in the furnace, neither must the

The fourth Book. 107

metall be wrought too hot, and the Marble whereon 'tis wrought must be of the hardest stone, and must be wetted, else the marble will break and scale.

To calcine Lead.

CHAP. LXII.

AT first Calcine Lead in a Kil as the potters do, and in great quantity. Usually in two days they calcine many a hundred pound of Lead. In calcining observe that the Kil be not too hot, but sufficiently heated onely, to keep the Lead in fusion, for otherwise 'twill not be calcin'd. When the Lead is melted it yields at the top a Yellowish matter. Then begin to draw forwards the calcined part with an Iron fit for the purpose, always spreading it in the internal extremity of the Kils bottom which should be of soft-stone, which will bear the fire. And the Kil must have a declivitie towards the mouth, which I pass by as a thing well known. When 'tis calcined once it must be put, and spread a second time in the Kil, to reverberate in a convenient heat, always stirring it with an Iron, and that for many hours till it come this second calcination to a good Yellow and be calcined. Then serce all in a fine serce, and what passeth not the serce recalcine it with new Lead. This is the way to calcine Lead in great quantities to make thereof store of Potters ware.

To make glass of Lead.

CHAP. LXIII.

TAke of this calcined Lead 15 pound and Crystall or *Rochetta* or *Polverine Fritt*, according as you would make the colours, 12 pound, mix them as well as possibly you can, put them in a pot, and at the end of 10 hours, cast them into water for by that time they will be all well melted, separate the Lead, and return the metall into the pot, which in 12 hours at most you shall have most fit to work.

The manner how to work the said glass

CHAP. LXIV.

TO work glass of Lead into divers drinking or other vessels, 'tis necessary before 'tis taken upon the Iron to be a little raised in the pot, and then take it out, and suffer it to cool a little, and then work it on the Marble being clear. At first let the Marble be well wetted with cold water that this glass may not draw away with it the Marble, and scale it; which it always doth when the marble is not wetted, and incorporates it into its self. This sticking of the marble makes a foul colour in the works. Wherefore continually wet the marble whiles this glass is wrought, otherwise all the fairness and beauty will be taken from it, Do thus as often as you take the metall out of the pot. This sort of glass is so tender, that if it be not cooled in the furnace, taken a little at a time, and held on the Irons, and the Marble continually wetted, 'tis impossible to work it. Which proceeds from the calcined Lead, which makes it most tender as a caudle.

Glass of Lead of a wonderful Emerald colour.

CHAP. LXV.

TAke of *Polverine Fritt* 20 pound, Lead calcined 16 pound, serce the two powders first by themselves, then when well mixed, put them in a pot not too hot, and at the end of 8 or 10 hours they will be melted then cast them into water and separate the lead. Put them a second time into the pot, and in 6 or 8 hours they will be melted, then cast them into water and separate the lead. This being twice done the metall will be freed from all the Lead, and all the unctuosity which calcined Lead and *Polverine* give it, and will acquire a most bright and shining colour, and in few hours 'twill run and come very clear, then give it brass thrice calcined (made as in Chap. 28.) six ounces

The fourth Book. 111

and therewith mix a peny weight of *Crocus Martis* made with Viniger, put in this mixture at six times, always mixing well the glass, and taking at each time the interval of saying the Creed. Let this glass settle an hour, then mix and take a proof thereof. When you like the colour let them incorporate 8 hours, then work them into drinking glasses, which will appear in a colour proper to the Emerald of the old Oriental rock, with natural shining and glittering.

Let this glass stand in a pot when sufficiently coloured, till it hath consumed all dregs, and till it be perfectly refined, and then 'twill be so like the natural Emerald that you can hardly know one from the other.

*Another wonderful Green Emerald
beyond all other Greens.*

CHAP. LXVI.

THis is made in every thing as the Emerald-green, in Chap. 65. but with this difference, that this onely takes six ounces of the powder of the *Caput mortuum* of *Vitriolum Veneris*, made as Chap. 131.132. and the other the same quantity of Brass prepared. This happily is the rarest Green that can be made any way whatsoever, which I have often made to my content.

The fourth Book. 113

Topaz colour in glass of Lead.

CHAP. LXVII.

TAke Crystal *Fritt* instead of *Polverine Fritt* 15 pound, Lead calcined 12 pound, mix and serce them both together, set them in the furnace not too hot, at the end of 8 hours, cast them into water, separate the Lead from the pot and glass, and repeat this twice, then hereto add half glass of a Gold Yellow colour, let them incorporate, and purifie for an Oriental Topaz.

A Sky or Sea-green in glass of Lead.

CHAP. LXVIII.

TAke Crystall *Fritt* 16 pound, Lead calcined 10 pound, mix and serce them well together, set them in the furnace, in

12 hours the stuff will be melted, cast both it and the pot into water, separate the lead, let them stand in the furnace 8 hours a second time, then cast them into water a second time, and separate the lead, put them in the furnace, and in 8 hours your metall will be most clear, then take of Brass calcined 4 ounces, of *Zaffer* prepared a quarter of an ounce, mix these well, and put in this mixture at 4 times to the glass of lead, and at the end of two hours mix well the glass and take a proof, then let the glass stand 10 hours, in which time the colours will be well incorporated, and the glass be very well perfected and be fit to be wrought in any works.

The colour of a Granat in glass of lead.

CHAP. LXIX.

MIX 20 pound of Crystall *Fritt* with 16 pound calcined lead, serce and put them into a pot, and to them of *Manganese* three ounces, of *Zaffer* half an ounce, both prepared, let them stand 12 hours cast them into water and separate lead, put them again into the furnace, and let them purifie 10 hours, then mix them, and take a proof, when the colour perfect, and of a fair Granat, work the glass as before.

Saphyre colour in glass of lead.

CHAP. LXX.

TAke 15 pound of Crystal *Fritt*, and lead calcined 12 pound, mix and serce them well together, then add to them two ounces of *Zaffer* and of *Manganese* a penny weight, both prepared, let them stand in the furnace 12 hours, cast them into water, and separate the lead, repeat this a second time, and you shall have the colour of an Oriental Saphyre, very beautiful and fair, with the mixture of a double Violet colour.

A Yellow Gold colour in glass of Lead.

CHAP. LXXI.

TAke of Crystall *Fritt*, and calcined lead of each 16 pound, mix and serce them well, and add to them of Brass thrice burned six ounces, *Crocus Martis* made with Vinegar 2 peny weight, put them well mixed in the furnace, let them stand 12 hours then cast them into water, separate the lead, set them in the furnace other 12 hours and in that time 'twill be clear, mix them and take a proof. If it wax green, give it a little *Crocus Martis* (which takes away the greeness) till it become a most fair Gold-Yellow colour, often made by me.

The colour of Lapis Lazuli.

CHAP. LXXII.

MElt the fairest Lattimo made, as in Chap. 55. with the whitest Crystall and most tender, in a pot, when 'tis

The fourth Book. 117

well melted, give it of Blew Painters Smalts, by little and little, and when the colour is good, let it stand in the fire two hours and make a proof, and when 'tis good let it stand 12 hours, mix them, and work them. If the metall rise put in a piece of leaf Gold to diminish the rising. This will be very like the natural *Lapis Lazuli*.

The way to Colour natural Crystal of a Viper colour, without melting it.

CHAP. LXXIII.

TAke natural Crystal of a good water, and very clear, free from Terrestriety, in several Pieces of divers Magnitudes, crude Antimony, Yellow Orpiment of each powdered two ounces, *sal Armoniac* one ounce, powder and mix well these three last, put this mixture in the bottom of a Chrysible that will bear the fire, and above this mixture the Crystalls in pieces, then cover this Chrysible with another, mouth to mouth, lute them well, and when they are dry, set them in coals, which kindle by little and little, and when they begin to fire, let them flame of themselves, and then they will smoak much, do this operation in a large Chimney, and avoid the dangerous and deadly fumes, when all these fumes are gone, let the Chrysible stand till the pot cool, and the fire go out of its self. Then unlute the Chrysibles, and take which out the pieces of Crystal, and those which are at the top will be tinged with a good Yellow colour, with a red Rubie, and Balass colours with fair spots, those which lay at the bottom upon the powder, and the residence, into the Wavie colour of a Viper. These pieces of Crystall may be wrought as Jewels at the wheel, and will receive a good polishing, lustre and shewing beauty, such as is in the Topaz, Rubie and Balass, if you give them foils suitable to their colour they make a fair shew, being set in Gold. Of these Crystalls you may colour a good quantity, since the charges and labour is but small, and in colouring a competent quantity there always come forth Some beautiful and fair.

The colour of a Balass, Rubie, Topaz, Opal, and Girasole in Natural Crystall.

CHAP. LXXIV.

TAke Orpiment of a Yellow-oringe-tawney colour in powder, Crystalline white Arsnic, of each two ounces, crude Antimony, *sal Armoniac*, of each one ounce, put this powder well mixed, into a very capacious Chrysible, and upon the powder, scales, and little pieces of Crystall, upon these small pieces larger and grosser pieces of Crystall of a fair water, without spots, if you would have a pleasing thing, let them be very large. And so fill the Chrysible, to which lute well another mouth to mouth, make a hole at the bottom of the uppermost of the bigness a Tare, that the air may draw thorow this hole the fumes of the materials which pass thorow the pieces of the Crystal.

Which tingeth the Crystall well, and better than when they pass thorow the joynt of the Chrysibles. When the lute is dryed, set them in the coals so that all the lower most, and half the uppermost be buried in the coals. Then kindle the fire by little and little, do as in the former and avoid the deadly fumes. The materials fume long, keep constantly a strong, and good fire. See you let not in any wind or cold air by windows or other places, for the pieces of Crystall being then hot, will become brittle will split and not be good. When the fire is gone out of it's self, unlute the Chrysible and you shall find the greatest part of the Crystall tinged with the true colours of Topaz, Chrysolite, Balass, Rubies, Girasole, and Opal with wonderful beauty. Those of the best colour may be wrought by the Jewellers, at the wheel and appear natural jewels, and the Crystall holds it's natural hardness which is great. At *Antwerp* I made good store, and amongst them, some of them were of a fair Opal colour and some of the Girasole. You may set them in Gold with foiles.

The fourth Book. 121

Be sure the Orpiment be good, for therein consisteth all the secret. If the work proceeds not well the first time, repeat it a second, and with practice you shall always do it without failing.

THE fifth Book. Wherein is shown the true way to make pasts for Emeralds, Topas, Chrysolite, Iacinth, Saphyre, Garnat, Egmarine, and other colours, of so much pleasantness and beauty, that they surpass the same natural stones in all things, except hardness.

With a new Chymical way (never yet used) to make the said pasts, taken out of Isaac Hollandus, *and far excelling all other pasts that have been hitherto made, both in beauty & colour.*

CHAP. LXXV.

I Believe there are few who desire, and seek not with all earnestness the knowledge

The fifth Book.

to imitate perfectly Emeralds, Topaz, &c. And in a manner all sorts of Jewels, which in colour, splendor, pleasantness and clearness, excepting hardness, excel the natural and Oriental, a thing very delightful and pleasant.

Wherefore in this present Book I describe the means to make them, with the circumstances and diligence necessary to be used. There is no doubt but he who shall set himself to the work with diligence, shall do much more than what I publish. The way lately practised by me, and taken from *Isaac Hollandus*, maketh pasts of incredible, and seemingly impossible beauty and perfection. 'Tis true the work is somewhat long and wearisome, yet I that have many times performed it, say 'tis very facil and plain, and (that which is above all) this way is true. Wherefore all pains, expences, and charges employed in such a like work, ought to seem small and light.

The fifth Book. 125

The way to prepare natural Crystall.

CHAP. LXXVI.

TAke natural Crystall the clearest that is possible, and put by firestones, Calcidonies, and Tarso and other hard stones, which though they Vitrifie, yet they make not so clear, lucid and shining stones as natural Crystal doth. The said stones used to make counterfeit Jewels, though they take polishing wonderfully, yet they always have something earthy, and obscure in them. But Crystall hath always something, that's aerial and transparent, and draweth near to the quality and nature of Jewels, especially those which are natural and Oriental. For they work far greater effects than the *Italian* or *Dutch*. Take then works made of Crystal, put them in Chrysibles covered at the top, set them in burning coals till they be both well heated & fired, then suddenly cast the Crystall into a very large pan, full of cold clear water. When the Crystal is cold recalcine, and heat, and cast it into fresh clean water, repeat this 12 times and be sure the ashes and filth be kept out of the Chrysible, and that the water be always very clean. When the Crystal is well calcin'd grind it to an impalpable powder as fine as the best wheaten flour, and that on a Porphyrie-stone, with a muller of the same, and then 'twill crumble and come to a flour, like refined Sugar. If you powder the Crystall in Brass mortars, with an Iron Pestle, you call make nothing there with but a green Emerald colour. Grind not above a spoonful at a time, and this grinding, and sercing must be often repeated, so long till no roughness remains, nor can be felt in the powder. For otherwise a past made thereof will give onely a durty and imperfect work, and will never be like natural Jewels. But if the Crystall be well ground 'twill make artificial gems, far excelling true natural stones in beauty, colour, clearness, splendor and polishing. Make a good quantity of this material that you may make all colours, for this is the prime material to make all Artificiall jewels, and shall be called hereafter Crystal prepared.

The fifth Book.   127

These rules often repeated by the Author take together.

1. That the whole be done cleanly, to this end lute all the pots wherein the Crystall is calcin'd, and wherein the pasts are baked with lute well dryed, before they be set to calcine or bake.

2. Take pots that will bear the fire.

3. Grind all on Porphyrie-stone, and not in metal, to a most impalpable powder, first singly, and then together.

4. Keep a just proportion in the dose of the Ingredients.

5. Mix the materials well before you bake them, and if the past be not sufficiently baked the first time, bake it again a second time in the potters furnace, and never break the pot till you see 'tis baked, for if you do the pasts will be foul, and full of blisters.

6. Leave the vacuity of a fingers thickness in the top of the pot, especially where 'tis said it swels much, or that you must put it in leasurely, lest the materials run out into the fire, or stick to the cover, and so make a foul colour.

The fifth Book.

How to make Oriental Emeralds.

CHAP. LXXVII.

TAke of Crystall prepared two ounces, ordinary Red-lead four ounces, mix and incorporate them well together, good *Verdigreas* two peny weight, *Crocus Martis* made with Vineger eight grains, Mix all well and set them in a potters furnace, in the hottest place thereof, as long as the fire lasts. To see whether the past be sufficiently baked and purified, clear and transparent, take onely off the cover made of lute, and if the past be pure and transparent to the bottom 'tis a sign 'tis baked enough. Otherwise relute, and bake it again, without breaking the pot, for then the past will be full of points and blisters. Let the fire be continued 24 hours with dry wood.

I set up a furnace at *Antwerp* a purpose, wherein I kept 20 pots of divers colours and with a fire in 24 hours melted and purified all of them, and to be the more secure, continue the fire six hours more and by this means the past will be very well baked, and little wood wasted. These pasts may be cut and wrought, in every thing as ordinary Jewels, they wholly receive the same polishing and lustre, and are set in Gold with foiles, as the other commonly are. This past is harder than ordinary.

To make a deeper Emerald colour.

CHAP. LXXVIII.

TAke of Crystall prepared an ounce, of Ordinary Red-lead six ounces and a half, mix them, and add, of *Verdegreas* about three peny weight, and 13 grains, of *Crocus Martis* made with vineger 10 grains. Proceed according to the rules, and you shall have a marvellous Emerald colour for small works, and to be set in Gold. This past must be baked more than ordinary, to wast that imperfection which Lead usually gives; this past is britler, but fairer than the former.

To make a more beautiful past for Emeralds.

CHAP. LXXIX.

TAke of Crystall prepared two ounces, Ordinary Red-lead seven ounces, mix and add to them of *Verdegreas* about ten grains to every ounce, and of *Crocus Martis* made with Vinegar ten grains onely at a time, mix them and proceed according to rule, and you shall have an Emerald past for small works, very fair and beautiful, but not hard, by reason of the plenty of lead. Wherefore bake it more than ordinary to take away the blackness, and unctuosity Lead naturally yields.

Another most fair Emerald.

CHAP. LXXX.

TAke of Crystall prepared two ounces, ordinary *Minium* six ounces, mix them, and add of good *Verdigeas* well ground 80 grains, mix and bake them for a most fair Oriental Emerald.

An Oriental Topaz.

CHAP. LXXXI.

TAke Crystall Prepared two ounces, ordinary *Minium* seven ounces, mix them, and bake them, for a marvelous Oriental Topaz, to work any kind of work you please.

An Oriental Chrysolite.

CHAP. LXXXII.

TAke of prepared Crystall two ounces, ordinary *Minium* eight ounces, mix them and add of *Crocus Martis* made with Vineger 12 grains, mix and bake them more than ordinary by reason of the great quantity of lead.

A Sky colour.

CHAP. LXXXIII.

TAke of Crystall prepared two ounces, ordinary *Minium* five ounces, mix them and add 21 grains of *Zaffer* prepared and ground, remix and bake them for a most beautiful Sky colour.

The fifth Book.

A Sky with a Violet colour.

CHAP. LXXXIV.

TAke of Crystall prepared two ounces, ordinary *Minium* four ounces and a half, mix them, and add about four grains of Painters Blew smalts, mix and bake them, this past will be a most fair Violet, and pleasant Sky-colour.

An Oriental Saphyre.

CHAP. LXXXV.

TAke of Crystall prepared two ounces, ordinary *Minium* six ounces, mix them well, & add of *Zaffer* prepared five grains, mix with the *Zaffer* of *Manganese* prepared seven grains, remix and bake them for an Oriental Saphyre, which will have a most beautiful Violet colour.

A deep coloured Oriental Saphyre.

CHAP. LXXXVI

TAke of Crystall prepared two ounces, ordinary *Minium* five ounces, of *Zaffer* prepared about 42 grains, add to the *Zaffer* of *Manganese* prepared eight grains, mix And bake them well, and they will make a deeper Oriental Saphyre, with a Violet colour of notable fairness.

An Oriental Garnat.

CHAP. LXXXVII.

TAke of Crystall prepared two ounces, ordinary *Minium* six ounces, mix them and add about 16 grains of *Manganese* prepared, wherewith mix three grains of *Zaffer* prepared, mix them all together, and bake them for a most fair & sightly Garnat.

A Deeper Oriental Garnat.

CHAP. LXXXVIII.

TAke of Crystall prepared two ounces, ordinary *Minium* five ounces and a half, of *Manganese* prepared 15 grains, wherewith mix four grains of *Zaffer* prepared, mix them all, this swells much, bake them for an Oriental Garnat, which hath in it a very fair Violet colour.

Another fair Garnat.

CHAP. LXXXIX.

TAke of Crystall prepared two ounces, ordinary *Minium* five ounces, mix them, and add 52 grains of *Manganese* prepared, wherewith mix six grains of *Zaffer* prepared, mix them all well and bake them for an Oriental Garnat fairer than any of the former.

Observations for Pasts and their Colours.

CHAP. XC.

OBserve, that the colours in the aforesaid pasts, may be made deeper, or lighter, according to the works they are to be used for, and also the will and humour of the maker. Small stones for rings, pendants, and ear-rings require a fuller, but greater stones a lighter colour. No rules can be herein given, though those given by me will give some light to the curious Artist, to whose judgement it must be left and who may find out and invent more and better colours. Besides I set down here onely colours from *Verdigreas, Zaffer* and *Manganese*. But a curious person and practical Chymist may extract a wonderful red from Gold and another fair Red from Iron, from Brass an exceeding fair Green, from Lead a Gold colour, from Silver a Sky-colour, and a much fairer from Granats of *Bohemia*, which are low-prized, for being small you may draw a tincture from them, as I have often done in *Flanders*, and this doth notable effects. The same may be done with Rubies, Saphyres and other Jewels. To write of these things would be a business too long for me, who speak so clearly in this present work. But the colours abovesaid will make pleasant works.

The way to make the abovesaid Pasts, and to imitate all sorts of Jewels, marvelous and never used.

CHAP. XCI.

THis way which I have taken from *Isaac Hollandus*, when I was in *Flanders*, to imitate Jewels, is not much used, and known perhaps to few persons, and though it be somewhat laborious, Yet by how much 'tis more laborious 'tis so much the fairer, and beautiful, than any made in any place whatsoever to this day, or at

The fifth Book.

least not shewn to me by any person. Wherefore I will shew the manner to make them, so clearly, and with so many circumstances and observations, that any one versed in Chymistry will be easily capable thereof and will do the work perfectly. Take Ceruss, otherwise call'd white lead, grind it very fine, and put it into a great glass body, and pour therein as much distil'd Vineger as will rise a palm above it. Observe that at first the vineger boils and swels wherefore put it in by little and little, till all the fury and noise is gone. Then set the Vineger on a hot furnace in sand, and evaporate away the eighth part of it, take it from the fire, and when the body is cold, decant leasurely the Vineger coloured enough, and impregnated with salt, which set aside in a glass vessel, then pour more fresh distil'd Vineger on the Ceruss, and evaporate and decant as before. Repeat this till you have extracted all the Salt from the Ceruss which is when the Vineger is coloured no more, nor hath any more taste of sweetness, which usually succeeds the sixth time. Then Filtre these coloured Vinegers mixt together, evaporate and dry them in a glass body, and the salt of Lead will be at the bottom of a white colour. Which set in sand in a glass body from the neck downwards well luted, but the mouth of the glass must be open, and the furnace heated for twenty four hours continuance. Then take the salt out of the receiver, powder it, and if it be Yellowish and not Red, set it twenty four hours in the fire, till it become as Red as *Cinaber*. Make a good fire, but not to melt it, for then all your labour and pains will be lost. Pour distilled Vineger on this Red-lead calcin'd, repeating this work as before till you have extracted all the salt from it, and separated all the dregs and terrestriety in whole or in part. Keep these coloured Vinegers in earthen pans glased six days, that all the terrestriety and imperfection may sink to the bottom. Then Filtre them, leaving the grosser part at the bottom as unprofitable, then cover the Vinegers in a glass body, and there will remain at the bottom a most white salt of lead, and sweet as Sugar, which dry well and dissolve in common water, let the solution stand six days in glased pans, separate the terrestriety at the bottom, Filtre and evaporate as before, and there will remain at the bottom of the glass a salt as white as snow and sweet as Sugar. Repeat this solution, filtration, and evaporation thrice. This salt is called *Saccharum Saturni*. Which put into a furnace into a body of glass in Sand, and at a temperate heat for many days, and it will appear calcin'd into a colour much redder than *Cinaber*, and as subtile and impalpable as the finest serced wheaten flour. This is call'd the true Sulphur of *Saturn* purified from all terrestriety, foulness, and blackness which *Saturn* had at first in its self. Now when you would make past for Emeralds, Saphyres, Garnats, Topaz, Chrysolite, Sky or any other colour, take the same materials, colours, quantities as abovesaid in the former receipts, except that instead of ordinary Red-lead, you shall take Sulphur *Saturni* working exactly in every thing as before. And you shall have Jewels of marvelous fairness in all colours, which very far surpass the forementioned, made with ordinary Red-lead. For this true Sulphur *Saturni* outgoeth all others, more than I write thereof, as I have seen, and often made it at *Antwerp*. Pasts made with this Sulphur, have not that unctuosity and Yellowness, as the other ordinary ones have, which in time shew their foulness, and the moisture and sweatiness which coming from within men much soil them, which happens not to those made with the said Sulphur. Wherefore think not that pains much, which will be well recompensed with the work and effect.

How to make very hard past of all colours.

CHAP. XCII.

TAke of prepared Crystall ten pound, salt of *Polverine* six pound, made as in Chap. 3. well dryed and ground on a Porphyrie, mix and serce them well together, Sulphur *Saturni* two pound, mix these three powders in earthen pans glased and clean, and with a little common water make with them a hard past, and of the past little cakes, each weighing three ounces, with a little hole in the midst of them, dry these in the sun, & then calcine them in the highest part of the potters furnace, or in other like fires, then powder and grind these cakes on a porphyrie, and serce them fine, then set them in pots in glass furnaces, to purifie three days, and cast them into water, and return them to the furnace for 15 days to purifie, that all the foulness and blisters may vanish, and the past remain most pure, like natural Jewels. And moreover this sort of purest glass will be tinged into all colours you desire. For example into an Emerald with Brass thrice calcin'd, as is done in ordinary glass, into a Sea-green, with Brass calcin'd to redness, made as in Chap. 24. and with *Zaffer* into a Topaz, into a Saphyre with *Manganese* and Zaffer, into Yellow with *Tartar* & *Manganese*, putting them in by parts, and into a Garnat also, with *Manganese* and *Zaffer* dividedly put in. And indeed this past imitates all Jewels and colours, and hath a wonderful shining and lustre, And in hardness too it imitates the jewels, Especially the Emerald, which will be made most fair and almost as hard as the true.

The sixth Book. Wherein is shown the way to make all the Gold-smiths Enamels, to Enamel upon Gold in divers colours, with rules, and the materials which colour, and what fires make those Enamels, with exact diligence and clearest demonstration possible.

*E**Namelling on Gold and other metalls is a fair and pleasing thing, and in it's self not only laborious, but necessary, since we see metalls adorned with Enamels of many colours make a fair and noble shew, enticing beyond measure the eyes of the beholders. And because 'tis one of the most principal, and a most necessary part of glass, and it appearing* *to me to be a thing grateful and pleasing to all, I set my self to describe many ways to make several sorts of Enamels, as a thing not vulgar, and belonging to this Art, and one of it's most noble Appurtenances. And that this work might not be deprived of a matter so pleasant, profitable and necessary, I have made this sixth Book for the delight and benefit of all.*

The Material wherewith all Enamels are made.

CHAP. XCIII.

Take of fine Lead 30 pound, of fine Tin 33 pound, Calcine them together in a Kil and serce them, Boil this Calx a little in clean water in clean earthen vessels, take it from the fire and decant off the water by inclination, which will carry with it the finer part of the Calx, put fresh water on the remainder then boil and decant as before, repeat this as long as the water carries off any Calx,

The sixth Book. 147

Recalcine the gross remaining Calx, & then draw off again the more subtile parts, as before. Then evaporate the waters which carried off the finer Calx at a gentle fire, especially at the last that the Calx may not be wasted, which will remain at the bottome much finer than the Ordinary. Take then of this fine Calx, of Crystal *Fritt* made with *Tarso*, ground and serced fine, of each 50 pound, of white salt of *Tartar* eight ounces, powder, serce and mix them well: Then put this stuff into a new earthen pot baked, giving it a fire for ten hours, then powder it and keep it in a dry covered place. Of this stuff are made all the Enamels of whatsoever colours. This shall be call'd the stuff for Enamels.

To avoid our Authors repetitions observe
1. The pots wherein Enamels are made must be glased with white glass and bear the fire.
2. Mix and incorporate well the colours and stuff for Enamels.
3. When the Enamel is refined, and the colour good, and well incorporated, take it from the fire with a pair of tonges for the Goldsmiths use.
4. The way to make Enamels is this,

powder, grind, and serce well the colours; and mix them first well one with another, and then with the stuff for Enamels, then set them in pots in the furnace, when they are all melted and incorporated cast them into water, and when dry set them in the furnace again to melt (which they soon do) make a proof, and if the colour be too high take out some of it and add more of the stuff for Enamels, and if too light add more of the colour at pleasure to your content, then take it out of the furnace.

A Milk-white Enamel.

CHAP. XCIV.

TAke of the stuff for Enamels six pound, of *Manganese* prepared 48 grains, cast it thrice into water when refined and melted.

An Enamel of a Turcois colour.

CHAP. XCV.

TAke of the stuff for Enamels six pound, melt refine and cast it into water, set it in the furnace again; when 'tis melted, and refined, put in of thrice calcin'd Brass three ounces, *Zaffer* prepared 96 grains, wherewith mix well 48 grains of *Manganese* prepared, mix them well and put them into the stuff at four times, mixing them well every time, let them incorporate, make a proof with your eye that you may know by the eye when the colours are good, as I have always done, because sometimes the powders colour more and sometimes less. Thus I did at *Pisa*, and by mine eye without weights coloured all sorts of glass.

Another Azure Enamel.

CHAP. XCVI.

TAke of the stuff for Enamels four pound, wherewith mix of *Zaffer* prepared two ounces, and mix with it at first of thrice calcin'd Brass 48 grains, mix these two powders well with the stuff for Enamels, set them in the furnace, and work according to the rules.

A Green Enamel.

CHAP. XCVII.

TAke of the stuff for Enamels four pound, put it in the furnace, and in ten or twelve hours 'twill be melted and refined, cast it into water, and put it again into the furnace in it's own pot, when 'tis refined, give it of Brass thrice calcin'd two ounces, wherewith mix of scales of Iron well ground two ounces, put them in at three times, mixing and incorporating them every time, and ever and anon see whether the colour please, when 'tis well take it from the fire.

Another Green Enamel.

CHAP. XCVIII.

TAke of the stuff for Enamels six pound, wherewith mix well *Ferretto* of *Spain* well ground three ounces, and mix with it 48 grains of *Crocus Martis*, put them into the furnace, &c. These furnaces are made from about four to six inches for all Enamels.

Another Green Enamel.

CHAP. XCIX.

TAke of the stuff for Enamels four pound, which in few hours will be refined, then cast it into water, and put it

152 *The sixth Book.*

again into the furnace, and let it refine, then add these two powders well mixed at three times, to wit of Brass thrice calcin'd two ounces, of *Crocus Martis* made with Vineger 48 grains, put them in the furnace, and when they are well incorporated, take them from the fire: This is a fair and good Enamel.

A Black Enamel.

CHAP. C.

TAke four pound of the stuff for Enamel, of *Zaffer* and *Manganese*, of each two ounces prepared, and well mixed, incorporate the stuff and colours, put them in the furnace in a large pot, and when refined cast them into water, then put them in the furnace again, and they will soon refine, and make a Velvet Black.

Another Black Enamel.

CHAP. CI.

TAke of the stuff for Enamels six pound, of *Zaffer* prepared, of *Crocus Martis* made with Vineger, of *Ferretto* of *Spain*, of each two ounces, grind and mix well together these three powders, with the stuff for Enamels, put them into the furnace, and when refined cast them into water, put them in the furnace again, and take the Enamel out when 'tis incorporated, and the colour pleaseth you. This is a fair Black.

Another Black Enamel.

CHAP. CII.

TAke of the stuff for Enamels four pound, *Tartar* four ounces, *Manganese* prepared two ounces, grind and mix these two powders well with the stuff for Enamels, set them in the furnace in a large pot, when melted and refined, cast them into water, and put them into the furnace again, let them refine. This is a most fair Velvet Black to Enamel upon metalls ordinarily.

A Red Enamel.

CHAP. CIII.

TO four pound of the stuff for Enamels; add two ounces of *Manganese* prepared, mix them well, and set them in the furnace in a large pot, when 'tis refined and melted cast them into water, set them again in the furnace, and when refined take them out.

This is a fair Purplish Enamel.

A Purplish Enamel.

CHAP. CIV.

TAke of the stuff for Enamels six pound, of *Manganese* prepared three ounces, of Brass thrice calcin'd six ounces, mix them all well together, set them in a furnace, and let them refine, then cast them into water, and put them into the same pot, let them boil, and when refined take them from the fire. 'Tis a good Enamel.

A Yellow Enamel.

CHAP. CV.

TAke of the stuff for Enamels 6 pound, of *Tartar* three ounces, of *Manganese* prepared 72 grains, grind and mix well these powders together, and then with the stuff for Enamels, put them into the furnace in a large pot, when refined cast them into

into water, and set them again in the furnace. This Enamel is of a fair Yellow to Enamel on Gold, where it shews not well, if you add not Enamels of other colours.

A Sky coloured Enamel.

CHAP. CVI.

TAke of the stuff for Enamels 4 pound Brass calcin'd to make a Sky colour, as in Chap. 23. of Sea-green made as in Chap. 22. of each two ounces, of *Zaffer* prepared 48 grains, mix first these powders well together, then with the stuff for Enamels, when they are refined cast them into water, return them into the pot, let them melt and refine. This is a very fair and beautiful Sky colour.

A Violet colour'd Enamel.

CHAP. CVII.

TAke six pound of the stuff for Enamels; of *Manganese* prepared three ounces, of thrice calcin'd Brass 48 grains, mix these two powders well together, then remix them with the stuff for Enamels, put them into the furnace, and cast them into water, put them into the furnace again, and do as before.

[Page 158 is blank]

The seventh Book. Wherein is shown the manner how to extract Yellow Lake for Painters, from Broom flowers, and all other colours with another way to extract Red Lake, Green, Azure, Purple, and all colours from all kindes of Herbs and Flowers.

And to make <u>Blew</u>, *Ultramarine, and Lake, from Cochneel, Brasill, and Madder for Painters, and also to colour discoloured* Turcoises; *another way to make a transparent Red, and a fair Red to Enamel upon Gold and Metalls, things neither Vulgar nor common.*

IN this Book is shown the way to extract all colours from Flowers and Herbs, for the use of Painters, which may serve also for glass; and Lakes of many colours, and *Ultramarine* from *Lapis Lazuli*, all which things though in particular useful for Painters, may notwithstanding serve to colour glass in the superficies, and also in the fire of the furnaces, such is the *Ultramarine*, and also the way to make a transparent Red in glass, which seems at this day to be wholly lost, as a thing not profitable, and to make a fair Red, to Enamel upon gold all materials in the Art of glass, and at this day much conceal'd, and known to few, and many other things which I judged meet to be put in this present work, which I believe will be acceptable to curious and ingenious Spirits.

A Yellow Lake to Paint, from Broom Flowers.

CHAP. CVIII.

TAke a Lee of Barillia, and of Lime, reasonable strong; and in this Lee, boil at a gentle fire fresh Broom Flowers, that the Lee may draw to it all the tincture the Flowers, which you shall know by taking the Flowers out and seeing them white, & the colour well taken out, and the Lee will be yellow like good *Trebian* wine: then take out these Flowers, and put this in earthen dishes (glased) to the fire, that the Lee may boil, and put into it so much Roch-Alum, that with the fire, all the Alum may be dissolved; then make a fire, and empty this Lee into a vessel of clean water, and it will give a Yellow colour at the bottom: let them settle, and decant off all the water, and again put upon them other fresh water, and decant it off; let the tincture first sink to the bottom, and do this so long, till you have taken out all the salts of the Lee and Alum from the tincture; observing that by how much the more you wash this tincture from the salt of the Lee and, Alum, by so much more will the tincture of the colour be fairer, and more, beautiful, washing it always with water to carry away the salt of the Lee and Alum, and at each time before you decant the water, let the Yellow tincture settle to the bottom. Repeat this process, until you perceive the water run out sweet and without saltness as 'twas first put in, and then at the bottom will remain a beautiful and fair Lake: which spread, when wet, upon pieces of white cloath, and dry it upon new baked Bricks in the shade, and you shall have a beautiful Lake of a Yellow colour, for Painters, and also for glass.

To extract Lake from wilde Poppies Flower-deluces, Red Roses, Red Violets, and from all sorts of Green Herbs.

CHAP. CIX.

GEt what quantity of the leaves of Flowers of what colour soever they be, let every colour be by it self, fair Green Herbs by themselves: proceed with them as in Chap. 108. and you shall have a Lake & true tincture & colour from every Flower, and Herb, which will be a fair and beautiful thing for Painters, and without doubt, worthy to be much esteem'd.

To extract a Lake, and colour to Paint;
from Orange Flowers, Red Poppies,
Flower-deluces, ordinary Violets,
Carnation and Red Roses, Borage and
Cabage Flowers, Gilli-Flowers, from all
Flowers whatsoever, and green from
Mallows, Pimpernells, and
all other Herbs.

CHAP. CX.

TAke of whatsoever Herb, or Flower, of whatsoever colour you will, which being bruised green upon a leaf of white Paper, tinges it with it's colour, these are good, but the Herbs and Flowers which do not so, are not good, then put into a glass body ordinary *Aqua vitæ*, the head must be as large as possible, and in the top thereof put the leaves of whatsoever Flower or Herbs, from which you would draw a tincture, then lute the joynts of the head, and thereto fit a receiver, then give a temperate heat, that the thinner parts of the *Aqua vitæ* ascending to the head, and falling upon the leaves and Flowers, may suck out the tincture, and distill thence into the receiver coloured Red, and full of the tincture of the Flowers, making all the subtile part of the *Aqua vitæ* to ascend so long as it comes coloured, and then distill this *Aqua vitæ* coloured in a glass vessel, which will come over white, and may serve at other times, and the tincture will remain at the bottom, which must not be dried too much but moderately, and thus you shall have the tincture or Lake from all Flowers, and Herbs, singular for Painters.

A Blew to make.

CHAP. CXI.

TAke Quick-silver two parts, flour of Brimstone three parts, *sal Armoniack* eight parts, grind them all upon a Porphery, and with the Quick-silver, put them in a glass with a long neck luted at the bottom

The seventh Book. 165

in sand, make a gentle fire till the moisture rise, then stop the mouth of the glass, and increase and continue the fire, as in sublimation, till the end, and you shall have a Blew, most fair and excellent.

How to colour natural Turcoises
discoloured.

CHAP. CXII.

PUt Turcoises discoloured, and become white, into a glass, pour upon them oil of Sweet Almonds, keep this glass upon temperate ashes and warm, and in two days at most the stones will have acquired a most beautiful colour.

A mixture to make sphears.

CHAP. CXIII.

TAke of Tin well purified and purged, three pound, Copper well purified one pound, melt these two metalls, first the Brass, then the Tin, and when they are well melted cast upon them six ounces of *Tartar* of Red wine onely burnt, and one ounce and a half of Salt-peter, then a quarter of an ounce of Alum, and two ounces of Arsnick, let them evaporate, then cast it into the form of a sphear, and you shall have a good material, the which you shall cause to be burnished and polish'd, which will shew well, and this is the mixture called steel to make sphears.

The manner how to colour within, Balls of glass, or other vessels of White glass, with all sorts of colours, which will imitate natural stones.

CHAP. CXIV.

TAke a Ball, or other sort of glass that is white & fair, & Isinglass which must be infused two days in common water, then put this infusion into a white pan with fair water, and boil it till all be well tempered, observing that the Isinglass will be very tender with much water, then take it from the fire, and when it is warm, put it into a Ball of glass, & turn the glass round, that the Isinglass may fasten and wet every where glass within, this being done let the moisture drain and run out, then have in order these colours powdered, to wit Red-lead, and casting it into the glass it will make the said colour stick (which will run in waves) cast it into many places through a tube, then throw in blew smalts making it stick in waves, within the ball. Then do the same with Verdigreas, then with Orpiment, next with Lake, all well ground, always casting the colours in many places in waves, which by means of the Isinglass which hath moistened the glass within, those powders will every where stick to the glass; and so shall you do with all colours. Then take Gesso well powdered and put enough thereof into the Ball, and suddainly turn it about, that it may stick every where to the glass within. Do this work nimbly whilst the moisture of the Isinglass glass lasteth, that the powder may stick well, then empty by the hole of the glass the Gesso which is within the Ball, which shall then appear of divers colours with a most fair appearance like the natural Toies of hard stones, and at last these colours (when the Isinglass is well dryed) stick so that afterwards they will not fall off; and alwayes their colour is most fair without. Fit to these Balls a foot of wood or of other stuff painted, and they are held for beauty before Cabinets, and for Merchants counting houses very fair.

Ultra-marine.

CHAP. CXV.

TAke fragments of *Lapis Lazuli*, found plentiful at *Venice* at a low price, let these fragments be well coloured with a fair Skie colour, lay aside those that are not coloured, calcine them well in a Chrysible, and so heated, cast them into cold water, repeat this twice, then grind them upon a Porphyrie, to an impalpable powder as fine as wheaten flour sifted.

 Take then three ounces of the Rosin of the Pine, Black Pitch, Mastick, new Wax, Turpentine, of each three ounces, Linseed Oyl, Frankincense, of each an ounce, dissolve them in a new earthen Pipkin at a gentle heat, stir and incorporate them with a Spatula, then cast them into cold water, that they may cleave in a lump for your need.

 Take for every pound of *Lapis Lazuli* ground as before, ten ounces of the aforesaid past of gums, which dissolve in a Pipkin at a gentle fire, and when it is well dissolved, cast in by little and little, the said powder of *Lapis Lazuli*, and incorporate it with the gum with a Spatula, I cast all the materials thus hot being incorporated suddenly into cold water, and bathing my hands with Linseed Oyl, made a round pastill hereof long and proportionally thick. Of these pastils you may make one or more according to the quantities of the materials, keep these pastils fifteen days in a great vessel full of cold water, changing the water every two days, then shall you boil in a Kettle common clean water, the pastils in clean and well glased earthen pans, and cast upon them warm water, and so leave them till the water is cold, the said water being emptied out, cast upon them new warm water, and when it is cold empty it out, putting in again warm water, and when it is cold, empty it out, putting in again warm water, repeat this so many times till the pastils be dissolved by the warmth of the water, then put in new warm water, and you shall see the water will be coloured of a Sky colour, decant the water into a pan well glased and cleansed. This casting on of warm water upon the pastils, must be repeated till it be no more coloured, but observe that the water be not over hot, but luke warm onely, for too much heat makes the *Ultra-marine* grow black. All these coloured waters strained into pans, have in them the unctuosity of the gums, therefore they must be left to settle 24 hours, that all the colour may sink to the bottom, then the water with it's unctuosity must be leasurely decanted off, put upon the pastils clear water, and then strain the cold water thorow a fine strainer, stirring the colour that it also may pass the strainer, and by this means a great part of the foulness and unctuosity will remain in the strainer, wash the strainer always with fair water. And with new water pass the *Ultramarine* thrice thorow the serce, washed every time, and then usually it's filthiness will remain in the strainer. Put the *Ultramarine* into clean pans, decant the water softly off, which dry of its self, and you shall have a most beautiful *Ultramarine*, as I have often made it at *Antwerp*. The quantity from a pound of *Lapis Lazuli* shall be more or less according as the stone is of a fuller and fairer colour. Then grind it to an impalpable powder on a Porphyrie (as is abovesaid) and 'twill arise most beautiful. If you take common Blew smalts ground on a Porphyrie to an impalpable powder, and incorporate it with the gumm pastils with the foresaid quantities, keeping them in digestion in cold water 15 days with *Lapis Lazuli*, and work thorowout as in *Lapis Lazuli*, you shall have a very fair and sightly Blew Bice, which will seem to be an *Ultramarine*. These Blews not onely serve for Painters, but to colour glass excellently.

A Lake from Cochineel for Painters.

CHAP. CXVI.

INfuse one pound of the shearings of the finest woollen Cloath in cold water a day, then press them well to take away the unctuosity the Wooll hath from the Skin, then Alum these shearings after this manner.

Take four ounces of Roch-alum, two

ounces of crude *Tartar* powdered, put them into a small pipkin with about three flagons of water, when it begins to boil put in the Flox, and let them boil half an hour at a gentle fire, then take them off to cool for six hours, after take out the Flox and wash them with fair water, Let them stand two hours, then press the water well from them, and let them dry.

A Magistery to extract the colour from Cochineel.

CHAP. CXVII.

COld water four gallons, wheaten bran four pound, Saline of the *Levant*, Fenugreck, of each a quarter of an ounce, them into a pipkin over the fire till the water become so hot one may just hold his hand in it, take them from the fire, cover the pipkin with a cloath, for twenty four hours, to preserve well the colour, then decant the Magistrie for use.

Put into a clean pipkin three gallons of

cold water, and one of the said Magistery when it boils, of Cochineel powdered, after this manner, in a Brass Mortar, powder and serce one ounce of Cochineel, so many times, till all pass the serce, at last take a little crude *Tartar*, pound it in the mortar, and 'twill take up all the tincture sticking to the bottom of the Mortar, and to the Pestle, mix this *Tartar* with the Cochineel serced, and as soon as the water in the pipkin boils put in the Cochineel, and let it colour the water whil'st you can say a *Miserere*.

    Then take the Flox Alumed as before which must first stand in a pan of cold water for half an hour, and when the water is well coloured, press well the water from the Flox, put it into a pipkin, and stir it about very often, with a little stick, that the Flox may be well tinged, let it stand half an hour over the fire that it may boil gently, then take the pipkin from the fire and take out the Flox, mixing it with a clean stick, put it into pans full of cold water, and in half an hour let all the water drain off, and put more cold water, let that drain, and press it well, and set it to dry in a place where no dust falls, spread it abroad that it may not become musty, and heat again. Take heed that the fire be always very gentle, for with too strong a fire the colour becomes Black. Then shall you make a Lee in this manner, to wit,

Take ashes of Vine branches, or of Willows, or of other soft wood, put them upon doubled Canvas, and pour gently on them cold water, let the water run into a pan, pour twice this strained liquour upon the ashes, and let the Lee settle 24 hours, that the ashes may sink to the bottom, and when 'tis pure and clear, decant it off into other pans, putting by the terrestriety which is not good.

Put the said coloured Flox, into a clean and cold pipkin, with the Lee, boil them at a most gentle fire, for so the Lee will be tinged with a Red colour, and will draw the tincture from the Flox, and at first take little Flox and press it well, and if the colour be discharged, take the pipkin from the fire, and this is a sign that the Lee hath drawn the tincture of the Cochineel from the Flox.

Hang an *Hyprocras* bag of Linnen, over a great and capacious pan, strain thorow this bag all the tincture from the pipkin, and let the Flox also go into the bag, when the Lee is drayned, press the bag where, the Flox are, that you may have all the tincture: Then wash the bag from the hairs of the Flox, turning them inside outwards, that they may come forth pure and clean.

Then take 12 ounces of Roch-alum powdered, put it into a great glass of cold water, let them stand till all the Alum is dissolved, then fitly place the said bag well washed from the hairs of the Flox betwixt two sticks in the air. The bag must be large at the mouth, and narrow at the bottom, sowed in the manner of a round pyramid, and under the bag set a clean pan, then cast all the Alum water into the pan where the tincture of Cochineel is, and you shall see the Alum water suddenly separate the tincture from the Cochineel like as a Coagulum doth. Then with a clean dish cast into the bag all the said tincture and Lee, which will run clear out of the bag, but the tincture will stick to the bag. And when all the water is well neer out, if haply any strain through somewhat coloured, pour it again into the bag and then this second time 'twill leave all the tincture

The seventh Book.

in the bag, and the Lee will then run white and discharged of tincture. Then take clean sticks, and therewith mix the tincture which sticks on the bag in gross pieces, and have in readiness new baked bricks, whereon spread little pieces of linnen, and on the linnen small pieces of Lake which you shall take out of the bag, let them dry well, spread them not too thick that they may soon dry, for when the Lake stands long wet it grows musty and makes a foul colour. Wherefore you may, when the brick hath sucked out much moisture take another new brick, and so you shall soon dry it. When 'tis dry take it from the linnen, and this is a good Lake for painters, which I have oftentimes made at *Pisa*. Observe, that if the colour be too deep, you must give it more Rock-alum, but if too light less Roch-alum, for so the colours are made according to your gust and will.

———————————————

———————————————

Lake of Brasil and Madder very fair.

CHAP. CXVIII.

IF you would make a Lake of these materials each of them by themselves, you shall do in every thing as is before said of Cochineel, colouring the water with one of these materials, but you shall not use so much Alum by an ounce as you did in Cochineel, for Cochineel hath it's tincture deeper than Brasil, & Madder have. Wherefore you shall give them their proportion which you shall find by practice. And also to one pound of Flox you shall use more Brasil or Madder, for they have not so great a tincture weight to weight as Cochineel hath. And in this manner you shall have a very fair Lake for Painters, and with less charge than that from Cochineel, and that from Madder in particular will arise most fair and very sightly.

Lake from Cochineel after another and more easie manner.

CHAP. CXIX.

IN this way invented by me at *Pisa*, you meet not with Flox nor Magisterie, nor Lee, nor dying the Wooll, nor so many things as go the former, which indeed is a very laborious way, though most true. But this way is most easie, and worketh the same effect, and 'tis this which followeth.

In a pottle of *Aqua vitæ* of the first running put one pound of Roch Alum well powdered, when it is all dissolved, put in an ounce of Cochineel powdered and sifted in every thing as before, put all this in a glass body with a long neck, and shake it well, and the *Aqua vitæ* will be wonderfully coloured, let them stand four days, then empty this stuff in to a clean earthen glased pan, then dissolve four ounces of Roch-alum in common water, cast this into the pan of *Aqua vitæ* coloured with

Cochineel, and put this into the *Hyppocras* bag, and so proceed throughout as in the 117 Chap. This is a most noble Lake from Cochineel, made with small pains, and in much greater quantity. All this was tryed at *Pisa*.

A transparent Red in Glass.

CHAP. CXX.

TAke *Manganese* ground to an impalpable powder, mix it with as much more refined Salt-peter, set it to the fire in an earthen pan to reverberate and calcine 24 hours, then take and wash it with common warm water from it's saltness the salt being separated, dry it, and it will be of a Red colour, hereto add it's weight of *sal Armoniack*, and grind them together on a Porphyrie, wet them with distill'd vinegar let them dry, then put them in a Retort which hath a large body and a long neck, give them a subliming fire in sand for 12 hours, then break the glass, and take all that is sublim'd to the neck, and body of the Retort, & mix it with the bottom & remaining residence, weigh them and add as much *sal Armoniack* as shall be wanting in this first sublimation, grind them all together on a Porphyrie, imbibing them with distilled Vinegar, then sublime them in a retort as before, and this sublimation is to be repeated after the same manner so long till the *Manganese* remain all at the bottom fusible.

This is the medicine that colours Crystal and past into a Red Diaphanous colour, and into a Rubie colour, there are used of this medicine 20 ounces, to one of Crystall or glass, but more or less may be used thereof according as the colour requires. The *Manganese* must be of the best from *Piemont*, to colour glass of a fair, and very sightly colour.

A Red as red as Blood.

CHAP. CXXI.

Put six pound of glass of Lead, common glass ten pound, into a pot glased with white glass, when the glass is boiled and refined, give it Copper calcined to redness according to discretion let them incorporate, mixing well the glass, then give it so much *Tartar* powdered that the glass may become as Red as blood, if it be not so much coloured, add Copper calcin'd to Redness, and *Tartar*, till it come to this colour.

The colour of a Balass.

CHAP. CXXII.

PUt Crystall *Fritt* in a pot into a furnace, cast it thrice into water, then tinge it with *Manganese* prepared into a clean purple, then take *Alumen Catinum* sifted fine, put in thereof so much as will make the glass become purple, and this you shall do eight times, and know that Alum makes the glass grow Yellow, and a little Reddish, but not blackish, and it always makes the *Manganese* flie away; and the last time that you add *Manganese*, give not the glass more Alum except the colour be too full, and so you shall have a most fair *Ballas* colour.

To extract the Anima Saturni *which serves for many things in Enamels and glass.*

CHAP. CXXIII.

PUt Litharge well ground into an earthen pan well glased pour upon it distilled Vinegar, which must be higher than it four fingers, let them stand till the Vinegar is coloured into a milkie colour, which it will suddenly be, decant off this coloured Vinegar, and put new upon the Litharge, repeat this work till the Vinegar becomes no more coloured. Then let these coloured Vinegars stand in earthen pans glased that the milkie substance of the Lead may sink to the bottom, decanting off the clear Vinegar, this milkie material is the *Anima Saturni*, to wit the most noble part, which serves for enamells, and glass in many things, and if this white stuff precipitate not well, cast upon it cold water, which is wont to make it fall to the bottom,

The seventh Book. 185

and when it doth not precipitate evaporate the Vinegars and waters, and the more subtile part remains at the bottom good for many things in this Art.

A fair Red to Enamel Gold.

CHAP. CXXIV.

TAke Crystall *Fritt* made in this manner, to to wit, salt of *Polverine* ten pound, white Tarso finely ground eight pound, make a solid past with this stuff, and water, and make thereof as it were small and thin wafers. Put these on earthen pans in a little furnace made in the fashion of a calcar, that they may be calcin'd with a good fire ten hours, and in defect thereof put them in the furnace, near the Occhio, for three or four days till they be well calcin'd. Take calcined Lead, and Tin prepared as in Chap. 93. *Tartar* of white wine calcin'd, of each two pound, mix them well together, and put them into a pot glased with white glass, let them melt, and refine well, then cast them into water, do this twice, then put them in the furnace, and when well refin'd in the pot give them of Copper calcin'd to Redness ten ounces. Let the colour purifie well, then give it *Crocus Martis* made with *Aqua-fortis*, putting it in by little and little, as you do with *Manganese*, then let it settle six hours, and see whether the colour be good, if not give it *Crocus* by little and little, till you have the desired colour.

A fair Red for Gold after another Manner.

CHAP. CXXV.

TAke Crystall *Fritt*, made as in Chap. 124. four pound, melt it in a clean pot glased, cast it, when refined, into water, and refine it again in the furnace, cast it into water a second time, and refine it again, then put in by little and little of calcin'd Lead and Tin purified, half an ounce at a time let the Calces incorporate, and when the glass becomes of an ash colour, put in no more Calces, For too much of them makes the colour white and not good. Let the glass refine with the calces, then put into the glass fine Red Lead two ounces, and when incorporated and refin'd well, cast them into the water, and set them in the furnace eight hours, then take of the Copper calcin'd to Redness, and of white crude *Tartar* of each half an ounce, put them and mix them well in the pot, then add of *Lapis Hæmatites*, wherewith the Cutlers burnish, and of fixed *Sulphur*, of each one *Drachm*, mix and incorporate these powders, and see if the colour be too deep, give it a little *Manganese*, which makes it lighter, and if it be too light a colour give it fixed Sulphur, and *Lapis Hæmatites*, and a little of Copper calcin'd to Redness, and a little *Tartar* of white wine with discretion, and do this till it come to the desired colour.

To fix Sulphur for the work abovesaid.

CHAP. CXXVI.

Boil Flowers of Brimstone in common oil an hour, take them from the fire, and cast upon them the strongest Vinegar, and the Sulphur will suddainly sink to the bottom, and the oyl will swim upon the Vinegar, empty the oyl and Vinegar, and put new oyl upon the Sulphur, repeat this thrice, and then you shall have a fixed Sulphur, for the work abovesaid.

Glass as Red as blood which may serve for the abovesaid fair Red.

CHAP. CXXVII.

MElt in a pot of glass of Lead six pound, Crystall *Fritt* ten pound, cast them when refined into water, put them again into the pot, when they are well refin'd give this glass four or six ounces of Copper calcin'd to Redness, let them boil, and refine well, then give them Red *Tartar* powdered, which incorporate with the glass, let them refine, and see if the colour please you, and is it be not heightned with the Copper, and *Tartar*, put it again to anneal till it come to be sufficiently Red, this is done to heighten the colour.

An approved way to make a fair Red Enamel for Gold.

CHAP. CXXVIII.

TAke of Crystall *Fritt*, boil it as in Chap. 124. six pound, refine it well in a glased pot, and give it fine Calx of Lead and Tin prepared, as in Chap. 113. four ounces at four times, when well refin'd and incorporated cast them into water, and then melt and refine them well again in the furnace, and give this glass at three times one ounce and a half of Copper calcin'd to redness, which makes the deep Red, mixing the glass well, and let this powder incorporate, and refine well in the glass, and within two hours give it *Crocus Martis* made as in Chap. 16. one ounce & a half at three times, let it mix and incorporate well in the glass three hours, then add six ounces of *Tartar* burn'd, with one ounce of the soot of the Chimny well vitrified, and with these powders mix half an ounce of the said *Crocus Martis*, put these powders well ground into the glass at four times, mixing them well, and interpose a little space between each time, for they make the glass swell and boil exceedingly, when all the powder is put in, let the glass refine three hours, then remix them, and take a proof, to wit, a little Bowl of glass, and scall'd it well, if it take a transparent Red, as blood, it's well, if not give it new *Tartar* burnt with soot, and *Crocus Martis*, by little and little, till it come to the desired colour, let the glass stand to settle, and an hour after you put in the powder, take another proof as before. This is good to Enamel, and proved often times at *Pisa*.

A transparent Red.

CHAP. CXXIX.

CAlcine Gold with *Aqua-regis*, many times, pouring the water upon it five or six times, then put this powder of Gold in earthen pans to calcine in the furnace till it become a red powder, which will be in many days, then this powder added in sufficient quantity, and by little and little to fine Crystall glass which hath been often cast into water, will make the transparent red of a Rubie as by experience is found.

The way to fix Sulphur for a Rose-Red to Enamel on Gold.

CHAP. CXXX.

TAke a strong Lee of Lime, and oaken ashes, boil sufficiently Sulphur in this Lee, which takes away a certain unctuous and combustible colour which Sulphur hath in it; by changing the Lee the Sulphur becomes white and incombustible and fixed, good to make this Rose-red for the Goldsmiths to Enamel upon Gold.

Vitriolum Veneris *which was began at the end of 31 Chap.*

CHAP. CXXXI.

SEt Chrysibles luted and covered in an open wind furnace with burning coals over them, let them stand two hours, and then at last let the furnace cool of it self then take out the Chrysibles, and you shall find the Copper calcin'd to a blackish colour, having an obscure purple, which powder, and serce well, then take a round vessel of baked earth plain at the bottom, which will bear the fire, set these pans in an open wind furnace, on iron bars set across, fill the pans with kindled coals, and put in the aforesaid calcin'd Brass, wherewith you have first mixed to every pound weight there of six ounces of common Brimstone powdred, & when the fire begins to heat the pans, and the Brimstone to flame and burn, continually stir the Copper with a long Iron having a hoock at the top, that it may not stick, nor cleave to the pans; continue this till all the Sulphur be burnt and smoak no more, then take the pans from the fire thus hot, and all the Copper, with an Iron ladle or like thing, powder it well in a Brass morter, and serce it, which will then be a black powder, proceed thrice with the same quantity of Copper and Brimstone in every thing as before. Observe, that at the third calcination you let the pans stand over the fire, so long that the Copper acquires a red Lion colour, then take it from the fire, and powder it in a Brass mortar, and you shall have the said colour to make the said Vitriol as we are about to say.

Vitriolum Veneris *(without Corrosives)* from which is extracted the true and lively Blew, *a thing marvellous.*

CHAP. CXXXII.

TO make then the *Vitriolum Veneris* above-said, take one or more very capacious Glass bodies, according to the quantity of the Copper calcin'd, and prepared, to wit, to a pound of Copper take a body which will hold six pints of water, put this common clean water into the body with calcin'd Copper into a sand furnace, give them a temperate fire for four hours, until of the six pints of water, there be evaporated about two, which is seen by the eye; let the furnace cool, and gently decant off the water into earthen pans glased, and the Copper which remains at the bottom put into pans in a furnace to evaporate all the moisture, and the water which is decanted into the pans will be coloured with a full and wonderous fair blew, let them stand thus in the pans two days to settle, and part of the Copper will sink to the bottom in a Red sub-stance, then Filtre the said water with usual linguets into glass vessels, and evaporate from the said Copper all the moisture, and with six ounces of Sulphur calcined, powder and serce it to a black powder, as in Chap. 131. and then as in the beginning of this pour in water and extract the Blew colour. Consider that in this work many pots will be broken, wherefore as often as the pots are broken or cleft take a new one, lest they break in the furnace, and all your labour be lost; when the humidity is evaporated put the same quantity of Sulphur powdered and serced, and do as before. The reason why the Copper is to be taken out whil'st it is hot, is, because then it is better separated from the pots, & it is impossible to separate it, if you suffer it to be cold, although you break the pots. Repeat this process not onely four but five or six times in every thing as before, Then the Copper will remain as a soft earth, and the better and most noble tincture of it will be in the Filtred waters, all which mixed together must be Filtred with the usual linguets, and the setlings and dregs may be cast away as unprofitable, then you shall have a most limpid water, and coloured with a most marvellous blew colour.

The way to extract Vitriol from the said colour'd Waters.

CHAP. CXXXIII.

SEt then a great glass body that will hold three Flasques of liquour in ashes or sand in the furnace, and with a temperate fire evaporate the said colour'd Waters, and neer to the furnace keep other glass bodies full of these colour'd waters, that they may be warm, and now and then fill the great body, which is in the sand with glass ladles, do this that the colour'd waters may be put in warm, for being put in cold they will make the great glass body break; evaporate the colour'd liquour from ten Flasques to two and a half or three, then these waters will be deep and full of tincture, which put in earthen glased pans in a cold and moist place for a night, and and you shall finde the Vitriol shot into points like crystals, which will appear like true Orientall Emeralds, decant off all the water that is in the pans, dry the Vitriol, and let it not stick to them, then evaporate half this water, which will yield you new Vitriol as before, Repeat this till you have gotten all the Vitriol. Put this Vitriol in a Retort well luted with a strong lute, See you put no more than one pound of Vitriol in a Retort, which must not be very large, and have a large and capacious receiver; make for 4 hours together a most temperate fire, for if it be too strong the moist and windy Spirits which first arise from this Vitriol, are so powerful, and arise with so great force, that no receiver is able to hold them; let the joynts also be very well luted. At last, make a strong fire when the dry Spirits begin to rise in a white form, continue the fire till the Receiver begins to wax clear, and to be quite cold, then make no more fire, and in twenty four hours let the joynts be unluted, and the liquor which is in the Receiver must be kept in glass very well sealed. This is the true lively Azure, with which marvellous things are done, as you may well perceive by it's smell, which is as powerful and sharp as any this day known in nature. Many things might be said, which are passed over as being not pertinent to the Art of glass, which happily you may judge upon better occasion; the feces then which remain at the bottom of the Retort will be black, which left some days in the air of themselves will take a pale blew, powder and mix this with *Zaffer*, and put it to Crystall metall as before, and with the said quantity will be made a marvellous Sea-green. Wherefore I have here set down the way to make this powder with much clearness, presupposing that I have not published an ordinary way to make it, but a true treasure of nature, and that to the content of noble and curious Spirits.

FINIS.

201

AN INDEX.

Lead *to calcine* 62. *To prepare* Crystall, 70, 76. Manganese 13. Sulphur *to fix* 126, 130. Vitriol *to make* Aqua-fortis 39. Tartar *to calcine* 37. *to burn* 41. *to extract* the salt 11. Zaffer *to prepare* 12.

To make Aqua-fortis 38. Aqua Regis 40. Crocus Martis *with* Sulphur 16. *with* Vineger 17. *with* Aqua-fortis 18. *with* Aqua Regis 19. Crystal *and* Crystalline Metall 9, 10. Frit 3, 8. Frits *of* Crystall 59.

Brass *to calcine* 20, 21. *for* Ferretto *of* Spain 14, 15. *to redness* 24. *thrice* 25, 28. Vitriolum Veneris 31, 131, 132, 133.

Glass *of* Lead 63. Saccharum Saturni 123.

Salt from Polverine Rochetta *and* Barillia 1. *a better way* 3. *from* Fern and other herbs 5, 6, 7.

Lakes 108, 109, 110, 118, *from* Cochineel 116, 117, 119, Sphears 113.

AN INDEX.

Turcoises *that have lost their colour* 112.
Ultramarine 115.
Blew 111.

Colours to make

Amethist in glass 48.
Balass in glass 122. *in Crystall* 74.
Black in glass 51, 52, 53. *in Enamels* 100, 101, 102.
Calcidony 42, 43, 44.
Chrysolite in Past 82.
Emerald in glass 32, 33. *in glass of Lead* 65, 66. *in Pasts* 77, 78, 79, 80,
Green in glass 33, 34, 35. *in Enamels* 97, 98, 99.
Girasole in Past 74.
Granat in glass 47. *in glass of Lead* 69. in Past 87, 88, 89.
Lapis Lazuli *in glass of Lead* 72.
Marble colour in glass 56.
Opal in Past 74.
Peach in glass 57.
Paste, observations on them 90, 91, 92.
Pearl colour in glass 60.
Purple Enamel 104.
Red deep in glass 58. *as blood* 121, 127. *in Enamels* 103.

AN INDEX.   203

Transparent in glass 120. *Rose red in glass* 120, 124, 125, 128.
Ruby in Past 74.
Sea-green in glass 22, 23, 26, 29, 30, 31, 131.
Saphyre in glass 49, 50. *in glass of Lead* 70. *in Pasts* 85, 86.
Sky colour in glass 23. *in glass of Lead* 68. *in Pasts* 83, 84.
In Enamels 106.
Topaz in glass of Lead 67. *in Past* 74, 81.
Turcois in glass 36. *in Enamels* 95, 96.
Viper colour in Crystall 73.
Violet Enamel 107.
White Enamel 94. *white call'd Lattimo in Glass* 54, 55.
Yellow in glass 46. *in Enamels* 105. *Gold Yellow in glass of Lead* 71. *Observations of Yellow in glass* 4.

Observations on the Epistle to the Reader.

COncerning *our Author, and this work, I find no other mention of him, than a bare naming him by* Garso *in his Book* della dottrina universale, *and by* <u>Borint</u>. de sufficientia, *Pag. 141. Neither could I ever find by strict inquiry that the other piece promised in the Epistle Dedicatory, and the Preface, concerning* Chymical *matters was ever published, neither have I read in any* Spagyrical *writers quotations drawn thence. Wherefore I may easily conclude, that it never came to light, and it is no wonder he found no incouragement by this Book, to put forth that, since this kind of learning most useful to mankind, was accounted, sordid, and below the speculation of men living in those times, who wholly busied their subtile wits, either in contemplations useless, or indeterminable most of whose notions were bare* λογομαχίαι. *But our most learned* Bacon, *a man of a most sublime, and, piercing intellect, in his incomparable* Novum organum, *hath fully confuted & shewed, the vanity & in efficacy of that other way, and hath more wisely substituted another more effective and operative, for the more solid promotion of Arts and Sciences. This way of useful learning hath been more experimently followed by some particular persons, but not universally throughout. But now 'tis like to make a considerable progress being designed by that most noble and honourable company of the Kings society at* Gresham College *which by the indulgence of His sacred Majesty, restored to his people, for the promotion of all virtuous undertakings, weekly convene to this very end and purpose, and daily bring in materials for this fair Edifice.*

One part of this design this present Book contains, wherein is set forth truly and plainly, the whole business of making and colouring glass, which from his youth our Author had learned of able and diligent persons, or what experience, or the fire had taught him, and in many he tells you the time and places of his tryal and invention, with all the circumstances thereunto belonging.

Art of Glass. Our English *word Glass is the same with the Dutch, and is deriv'd from the* Latine, Glastum, *which by removing the last syllable, is plainly Glass; now it appears that* Glastum *was called* Vitrum, *by* Cæsar *in his* Commentaries, lib. 5. *Where he saith,* omnes Britanni se vitro inficiunt *all the* Britans *colour themselves with Glass, &* Mela, lib. 3. cap. 6. Britanni vitro corpora infecti, *and* Vitruvius, *wooll died with Glass, for so the learned* Turnebus *restores these places, where 'twas anciently read* Ultrum *for* Vitrum; *but that* Vitrum *is* Isatis, *appears by these words of* Vitruvius, *they colour for want of* Indico Chalk *from* Selinutia Vitro, *with Glass, which the* Greeks *call* Isatis, *as also by a* Treatise *of* Apuleius de herbis, *not published, but is in the hands of Doctor* Merrick Causabon, *larger and more correct than those that are published, he thus,* Herbam Isatis alii Aogigneme prophetæ Apesion Itali alutam alii herbam vitrum, *which is to be written,* Isatis alii Angionen Prophetæ Arosion Itali rutam alii herbam vitrum. Salmatius *ever <u>falsly</u> puts* Guastum *for* Glastum, *because the* Britans *continually call it* Guadum, *The which call a Blew colour Glass. And* Pliny, lib.22. chap. 1. *witnesseth the same in these words,* simile plantagini Glastum in Gallia Vocatur quo

Britannorum conjuges nurusque toto corpore oblitæ quibusdam in sacris nudae incedunt. *The British women cover their bodies with* Glastum, *& in some Festivals go naked. And* Cambden *in his* Britannia, *this is the herb we term* Woad, *and it gives a Blewish colour, which the* Britans *at this day call* Glass. *The reason why* Glastum *acquired this name* Vitrum, *or Glass, might be, because all glass hath Naturally (as this Author and experience teacheth) somewhat of blewishness in it.* Vitrum *comes from* Visum *as* Aratrum *and* Rustrum *come from* Aratum *&* rutum, *the last syllable being changed into* trum, *so* <u>Isodurus</u>, lib. 16. cap. 15. Quod visui perspicuitate transluceat, *because it is transparent to the sight for in other metals, whatsoever is contained within is hid, but in Glass all liquors, and things within appear the same as without, hence it is that many transparent bodies are call'd* Vitrea, *as the humour of the eye, the Sea, Rivers, waters by Physicians,* Horace, Ovid, *and* Boeth, *and* Apuleius *of a spring.*

Glass is one of the fruits of the fire, which is most true, for it is a thing wholy of Art, not of Nature, and not to be produced without strong fires. I have heard, a singular

Artist merrily to this purpose say, that their profession would be the last in the world: for when God should, consume with fire the Universe, that then all things therein would vitrifie and turn to glass. Which would be true upon supposition of a proportionable mixture of fit Salts, and Sand or Stones.

'Tis much like all sort of mineral or middle mineral. *I find Authors differ much about referring Glass to it's Species,* Agricola lib. 12. de Metallis, *maketh it a concrete juyce,* Vincent Belluacensis, lib. 11. *calls it a stone,* Fallopius *reckons it amongst the* Media mineralia, *and the work men, when it is a state of fusion call it metall. But to me it seems neither of these, which this generall Argument sufficiently evinceth, that all the above mentioned are natural concretes, but Glass is a compound made by Art, a product of fire, and never found in the bowels of the Earth as all the others are.*

Wherefore as factitious words of Art are excluded out of the predicaments by the Logicians, *so is Glass to be excluded out of the former Species. Neither is it more to be call'd a Metall, concrete juyce, than Beer or Malt, Barley, or Lime, a Stone, or Brick, Earth, &c. But to this argument* Fallop, *thus replies, by:*

asking of what Glass we speak, whether of that which is in it's own Mine, and it's own stone, or else of true Glass, and now extracted from the stone? if of this purified, he saith 'tis no more Artificial, than a metall is extracted from it's Mineral, and purified. But if we understand it of that which is the first stone then he saith that as metall in the Mine and proper stone, so glass having it's existence in the stone, whence 'tis educed, is natural. To whom I answer, that Glass is never found that form in any Mine, but onely Sand, and Stones which are the Materials of it. But of Metalls 'tis far otherwise, which nature hath perfectly formed into a certain Species in Proper veins, though sometimes they are by the fire forced out of the veins, and Earth or stones wherein they in smaller particles and Atoms lay hid. And with this difference too, that fire onely produceth or rather discovereth Metall by it's innate energie of separating heterogeneous bodies and congregating homogeneous: But in Glass 'tis far otherwise, for that is made by uniting and mixing different parts of salt and sand. Which Fallopius *to admiration denies saying, that 'tis false that Glass is made of Ashes, and he adds, that although Glass-men add ashes brought from* Alexandria, *or from other places, yet he saith that ashes is added instead of Nitre which the Antients used, that they might more easily extract Glass from the Metalline stone. Yet we may not say that ashes is mixed with the Metall to make Glass, but that 'tis onely put into the furnaces where Glass is melted, that Glass may be more easily educed from the smallest and inmost particles of the Glass-stone, that is, of it's proper Metall; so far he. But this strange opinion is easily confuted; for if Glass were extracted from the stones onely, then the weight of the Metall must needs be far less than the stones alone, but in truth the weight of the Metall far surpasseth that weight, for 100 weight of Sand yields above 150 of Metall; besides, the Salts composing Glass are the most fixed salts, which the fire cannot raise with the most vehement heat. Again in old windows of French Glass, in that part which lies towards the air, you may manifestly discern, nay, pick out pieces of salt, easily discovering their nature to the tast, furthermore in the finest Glasses, wherein the salt is most purified, and in a greater proportion of salt to the sand, you shall find that such Glasses standing long in subterraneous and moist places will fall to pieces, the union of the salt and sand decaying. And this is the reason of that saying, that Venice Glasses will break with poison, which is true of some Mineral, but not of Vegetable or animal poyson. All which manifestly evince that salt remains in the Glass in specie. Add hereunto that experiment of* Helmont, Cap. de terra, *who thus saith,* Si vitri pollinem pluri alkali quis colliquaverit ac humido loco exposuerit, reperiet mox totum vitrum resolvi in aquam, cui si affundatur Chrysulea, addito quantum saturando alkali suffecerit inveniet statim in fundo arenam sedere eodem pondere quae prius faciendo vitro aptabatur. *If you melt fine flour of Glass with good store of* Sandever, *and set them in a moist place, you shall soon find, all the Glass resolved into water, whereunto if you pour as much* Aqua-fortis, *as will suffice to saturate the* Sandever, *you shall find the sand presently settle to the bottom in the same weight which was put in at first.*

And, in this experiment the salt is imbibed, and taken up by the Sandever, *and* Aqua-Regis, *and so the component parts analysed into their former principles, which were before confused in the compound.*

A second general argument is this, that though the said concrete juyces stones and

Glass, may have fusion in the fire yet neither all stones, nor all concrete juyces, Metalls, nor Semimetalls have fusion, such are Talc *and* English Spaud, sal Armoniack, Tincal, &c. *Reckoned Amongst concrete juyces; nor* Diamonds, Cats-eyes, Agate, Jaspers, *nor most other pretious stones, nor Marble; Nor many other stones wherewith the inside of these furnaces are built. Neither can* Mercurie *amongst metalls be said to melt, nor amongst the middle Minerals* Orpiment; *and though most of them have fusion, yet none of them have ductilitie, but Metalls onely, and they onely too, when they have received a great degree of cold; for when they are red hot the particles of them stick not together, nor are so Tenacious as Glass is, which onely whilst it is red hot, will with small force of the breath receive any fashion or figure, and by blowing form a cavity, none whereof any of the aforesaid bodies will do; besides metall poured out, when melted, will run into many small globuli, or pieces, but glass sticks together in a lump even in the furnace it self, when the pots are broken. And this quality of ductility, and tenacity I make to be the essential difference of glass from all other bodies; nay from all other substances, which have gotten the name* of glass, as Vitrum Antimonii, Moscovie glass, and, bricks or other stones vitrified, neither whereof will bear this tryal. Which rather have their denomination from their transparency, (as Vitriolum too hath a Vitro) than from their intrinsecal nature and properties. But to shorten this comparison, I shall here set down the proprieties of glass, whereby any one may easily difference it from all other bodies,

1   'Tis a concrete of salt and sand or stones
2   'Tis Artificial.
3   It melts in a strong fire.
3   When melted 'tis tenacious and sticks together,
4   It wasts not nor consumes in the fire.
5   'Tis the last effect of the fire.
6   When melted it cleaves to Iron, &c.
7   'Tis ductile whilst red hot, and fashionable into any form, but not malleable, and may be blown into a hollowness.
8   Breaks being thin without annealing.
9   'Tis friable when cold, which Made our proverb, As britle as glass.
10   'Tis diaphanous either hot or cold.

Epistle to the Reader. 215

11 *'Tis flexible and hath in threeds* motum restitutionis.
12 *Cold and wet disunites and, breaks it, especially if the liquors be saltish, and the glass suddainly heated.*
13 *It onely receives sculpture, and cutting, from a* Diamond *or* Emery *stone.*
14 *'Tis both coloured and made Diaphanous as pretious stones.*
15 Aqua fortis, Aqua Regis, *and* Mercurie *dissolve it not as they do Metalls.*
15 *Acid juyces nor any other thing extract neither colour, tast, or any other quality from it.*
16 *It receives polishing.*
17 *It loseth nor weight, nor substance with the longest and most frequent use.*
18 *Gives fusion to other Metalls and softens them.*
19 *Receives all variety of colours made of Metalls both externally and internally, and therefore more fit for Painting than any other thing.*
20 *'Tis the most plyable and fashionable thing in the world, and best retains the form given.*
21 *It may be melted but 'twill never be calcined.*

22 *An open glass fill'd with water in the Summer will gather drops of water on the outside, so far as the water reacheth, and a mans breath blown upon will manifestly moisten it.*
23 *Little balls as big as a Nut fill'd with* Mercury, *or water, or any liquor, and thrown into the fire, as also drops of green glass broken fly assunder with a very loud & most sharp noise.*
24 *Wine Beer nor other liquors will make them musty, nor change their colour nor rust them.*
25 *It may be cemented as Stones and Metals.*
26 *A drinking Glass fill'd, in part with water (being rub'd on the brim with the finger* wetted *yields Musical notes, higher or lower according as tis more or less full, and makes liquor frisk and leap.*

Antiquity of Glass.

C*oncerning the* Antiquity of Glass, *our Author here fetcheth it from* Job Chap. 28 .v.17. *Who in this* Chapter *from v. 15 to the 20th compares wisdom to the choicest things; and in this* 17th v. *saith,* Gold and Glass shall not be equalled to it. *So our Author from the* Vulgar Latine translation, *the* Septuagint, Hierom, Senes, Elias in Nomenclatore. Hieron. Pineda, Biblia Tigurina, & Syriac, but Iacinth in the Arabick translation.

Crystall, Chaldee, Santes, Arias Montanus,

Forsterus. *The* Hebrews *whom* Nicetas *follows, and the* King of Spains *edition, and so the* English translation.

A stone more pretious than gold, *as* Pagninus from Rabbi Levi Kimhi.

A Looking glass, *as the* Thargum *renders it; perhaps because in that time or age* Looking glasses *were first invented and highly valued, being made of precious materials, and so* Muncer *reads it.*

Glass of Crystall, Vitrum Crystallinum, Complutensis.

A Beril, *as* Vatablus.

A Diamond, Rabbi Abraham, Rabbi Mardocai, Pagnin, Cajetan, *the* Italian, Spanish, French, High *and* Low Dutch.

A Pyropus *or* Carbuncle, or some such neat and precious Gemm, as others, so Pineda: *But both those are the same name of one stone which the* Antients *gave to such a gemm as would shine by night, but there's none such in nature, & the later writers take the* Ruby *for it.*

The reason of this difference in the translators, is, because the Original word Zechuchih *comes from the root* Zacac *which signifies to* purifie, *to* cleanse, *to* shine, *to be* white, *and* transparent. *The same word is applied to* Frankincense, Exod. 30. 34. *and is rendred, by the* Septuagint, Pellucid, *Hence 'tis manifest why so many rendrings of the text, since the word in general signifieth onely what's transparent and beautiful, therefore the translators might apply the word to any thing which was of price and value, for so the text requires, and transparent too, for so the word requires. But it seems to be neither* Diamond, Carbuncle, *nor* Iacinth, *for those are mentioned in Aarons Brest-plate, Exod. 28. and this word here not to be found in that Chapter. Nor* Glass *nor* Crystal, *because 'twould seem incongruous, that those of so mean a value should be brought into comparison, the former being made of Materials very common, and the latter could not but be vulgar. Besides, 'tis probable this word subjoyned to* Gold, *was added after it for amplification. Add hereunto, that Glass is no where mentioned in the* Old Testament, *though frequently in the* New *by* S. Paul, S. James *and in the* Revelation. *And indeed who can imagine that a thing fit for so many illustrations, and comparisons, and of so common use, could be passed by in silence, if known, by the Scripture so full of elegancies in this kind? And therefore I judge it meet to keep the general word, and not to confine the sense to one pretious and*

Epistle to the Reader. 219

transparent stone, or thing, but to extend it wider to all things that have those two properties in them, But too much of this in messe aliena.

Aristophanes *seems to be the first that mentions this word* υαλος, *now rendered* Glass; *for in* Nubibus, Act. 2. Scen. 1. *he brings in* Sthrepsiades *abusing* Socrates, *and teaching him a new way to pay old debts,* viz. by placing a fair transparent stone sold by the Druggists, and from which they strike fire, betwixt the Sun and the accusation brought in writing against him, for the Sun would soon melt away the letters of the accusation, *which stone* Socrates readily call'd ὐαλος. *Whereon the* Scholiast *thus,* Druggists sold pretious stones as well as Medicaments. *And that the Antients call'd* Χρίον *(the same with* χρυςαλος*)* Crystall. That Homer *knew not the name, and that with him and the* Antients, *the word* Electrum *was used, the Scholiast there testifieth, though he himself clearly describes our Glass in these words.* We properly call that Glass which being melted by fire from a certain herb burnt to prepare certain vessels. Hesichius *hath not the word* υαλος *in this sense, but* Hyalen, Hyalon, Hyaloen, shining *and* Diaphanous.

The Etymologist, *hath it in this sense and fetcheth the* Etymon *from* υειν, *to rain, from the likeness it hath to ice (which is congeled rain or water) in consistence and* Diaphaneity, *and in this sense, as some Glass from glacies ice.* Aristotle *hath two Problems of Glass, first,* Why we see through it, Sect. 2. 61. *secondly,* Why it cannot be bended. *Now if these Problems were* Aristotels *(as learned men doubt whether they are or no) then this seems to be the most* Antient *piece of* Antiquity *for Glass. For neither in the* Antient Greek *Poets nor Orators shall you find an, mention of Glass, though a thing so fit for their purpose, as was abovesaid. And note the ambiguity of the word* υαλος, *for Crystal was so call'd as the* Scholiast *above, and* Hugo Grotius, *and these names are wont to be mixed by reason of the likeness of the things, and* Gorræus *saith, that, a certain kind of* Yellow Amber, *and transparent as Glass, was call'd by some* Hyalus. *The first then amongst the Greeks, that without question have mentioned Glass are* Alex. Aphrod. *who thus saith,* As the Floridness of a colour is seen through Glass, *and yet more clearly,* lib. 1. Probl. Glasses in the winter in vehement and sudden heat coming upon them, break,

Epistle to the Reader. 221

and again, to break the Body of the Glass. *And* Lucian *mentions very large drinking vessels of Glass. And* Plutarch *in his Symposiack, saith,* that fire of Tamarisk wood is fitted to form Glass.

That the Egyptians *were skilful in this Art, appears by* Flavius Vopiscus, *quoted by* Marcel. Donatus, *in these words,* Alexandria a City rich, fruitful, wherein no body lives idle, some Blow glass, others make Paper, *&c. Though* Kircher *in his* Oedipus, *writing of the* Egyptian Arts, *mentions not this.*

Lucretius *amongst the* Latine *Poets, is the first I find mention Glass, whose Verses I shall add, because they give his account of it's transparency.*

_____ *nisi recta foramina tranant Qualia sunt vitri,* l.4. 602, 603. and again, *Atque aliud per ligna aliud transire per Auram, Argentoq; foras, aliud vitroq; meare.* l. 6. v. 989, 990.

But downwards all the other Poets.

This Art *was unknown to* America *and all* Asia, *except* Sidon, *and* China, *who of late, have learned to make it very perspicuous of Rice, but very brittle, and therefore not to be compared as yet with ours, though it come neer it.* Atlas Cinicus, pag. 6.

But to decide this controversie, 'tis manifest that Glass could not be unknown to the Antients, and must needs be as Antient as Potterie it self, or the Art *of making* Bricks, *for scarcely can a Kill of* Bricks *be burnt, or a Batch of* Pottery *ware be made, but some of the* Bricks, *and ware will be at least superficially turn'd to Glass. And therefore without doubt 'twas known at the building of the Tower of* Babel, *and as long before as that* Art *was used, and likewise by the* Egyptians: *for when the children of* Israel *were in captivity, we read that making of bricks was a great part of their bondage. And of this nature mast that Fossil Glass be, whereof* Ferant Imperatus, lib. 25. cap. 7. *thus saith, Glass like to the* Artificial *is found under the earth in places where great fires have been, neither whereof struck yield sparks of fire. Other Glasses are found in round clots like fire-stones, shining in the breaking, and transparent with greeness, which in shew resemble* Colophonia, *and these struck sparkle like fire-stones. From which notwithstanding they are different as well in their Vegetation proper to firestones, as also in shining, and much quicker melting, proper to Glass. of these said Glasses some are brittle, others solid, the brittle or crumbling put in the fire, swell, and take the shape of white pumice-stone, and afterwards the shining of Artificial Glass: But those which are continued and solid, by a small change from the fire, pass from blackness to white* Artificial Glass, *This Fossil Glass is wrought by the* Americans *to make holes, and cut instead of Iron. So far he. And, happily of this sort of Glass, was a piece thereof, which I lighted on at S.* Albans, *an antient garrison of the* Romans, *which I struck off from a Roman Brick, 'tis of the same colour and substance with what appears in ours at this day.*

And no doubt but this Glass was more frequent in their Brick than ours, for they tempered their earth two years together, and so it wrought more firm, and close; besides, they burnt them better. And this vitrification of earths made into bricks, is not onely at the first burning of them, but also as Imperatus *observes might be from great fires to wit such as are in lime-Kils, and* Potters *Kils, such as were most* Antient *in* Asia *and* Africa, *for in those the*

Bricks *usually Vitrifie. But I have not heard nor seen any of them Vitrified in the firing of houses built therewith. For it seems that onely a fire made with layers of dried crude* Bricks *burnt in the fire, can produce this effect, or else by the way of Reverberation in furnaces where most vehement close & continued fires are made.*

This Glass lay long in the earth, though Helmont *affirms that Glass there dissolves, putrifies and, turns to water, in few years. Which though true in our finer Crystal, as to the saline part, yet seems not so of Glass in general.*

As for the way mentioned by our Author *found out by Merchants, it seems not very credible, since the continual burning of* Kali *in* Spain *and,* Egypt, *for* Barillia *and,* Polverine, *and of* Kelp *and, other Materials for green Glass with us, in greater quantities than the said Merchants did to dress their provision, and, consequently a stronger and more lasting heat raised thereby, did never produce Glass in any place or time whatsoever, nay the strong and close heat of the calcar cannot produce it; Perhaps those that refine Metalls from the* Ore, *whereof* Tubalcain *was the inventor or* Antient Chymists, *could not but <u>both</u> in their furnaces and from their Metalls long wrought upon by the fire, observe Glass also.*

Epistle to the Reader.

Amongst those Chymists, the most antient seem to be Egyptian *Princes, who all from* Hermes Trimegistus *downwards professed this art, indeavouring at an universal Medicine, but not the supposed transmutation of Metalls as* Kircher *in his* Alchymia Hieroglyph, *affirms. Now this attempt could not be without great fires and furnaces, which must at some time or other run into Glass, and, their materials also must do the like.*

So that it plainly appears by what hath been said, Glass must be known from great Antiquity, But the art of making and working Glass seems by what hath been said to be of later invention, and the first place mentioned for the making of it to be Sidon *in* Syria, *which was enobled for glass-houses and making of Glass,* Plin. l. 36. cap.26. *And that Glass was made in the time of* Tiberius *(the first we read of amongst the* Romans*) 'tis apparent by the history of the man whom* Plin. *relates he put to death for making Glass malleable, of which hereafter.*

Of the use of Glass

IN Domestick affairs it makes drinking vessels, infinite in fashion, colour, largeness, the Romer *for* Rhenish *wine, for Sack, Claret,*

Beer, plain, moulded, coloured. in whole or in part; Bottles and other vessels to keep Wine, Beer, Spirits, Oyls, Powders, wherein you may see their Fermentations, separations, and whatsoever other changes nature in time worketh in any liquours, the clearness and goodness of them. Besides dishes to keep and to serve up sweetmeats, glasses to measure time, sleekstones for Linnen, Ornaments for studies, and Presses, Windows to keep us warm and dry, and to admit Light into our dwellings, which passing through coloured Glass, it tingeth with the same colour whatsoever lyeth in opposition to the Sun. And lastly Looking glasses, the delight and business of Narcissus *and his followers.*

In Physick, Convex Spectacles for aged persons, and Concave Glasses for such as are Purblind, and cannot see unless the object be placed neer their eyes, contrary to the former, besides Cupping glasses, Urinals, and to draw Womens-breasts, in preserving the eyes of Engravers, and Jewellers, when they work some small and accurate work, and also for delight in Magnifying, to make artificial eyes, for Ornament, Diminishing, Dilating, Lengthening and Multiplying Objects, and variously changing their figure, and Situation, and by various placing of them to work astonishment and fear *in the vulgar beholder, as you may see in* Schottus Opticks, Catopticks, Catoptrocausticks, Catoptrographicks, Dioptricks, Telescopicks, *who hath there collected out of* Kircher, Porta, *and other Authors whatsoever is rare and admirable.*

In Astronomie, what strange wonders and Discoveries have those Telescopes, *invented by* Galileo *or* Scheiner *(for they both contend about it) and since exceedingly promoted by Sir* Paul Neal *an honour to the* English *Gentry, and the most learned* Hugenius, *the incomparable* Hevelius, *and by* Eustachio Divini *at* Rome? *The use whereof hath made the Doctrine of the Heavens very clear, daily detecting new Stars and new Worlds, things wholly unknown to the ancients, besides their use by Sea and Land, for Sea-men, Soldiers and all other Persons, to discern, and distinguish things at distance. Hereunto add that excellent Sphear of Glass, whereof* Claudian *writ that witty Epigram, which take* Englished *by M.* Randolph.

Jove saw the Heavens fram'd in a little glass,
and laughing, to the Gods these words did pass,

Comes the power of mortal cares so far?
In brittle Orbs, my labours acted are,
The statutes of the Poles, the Fates of things,
The laws of Gods the *Syracusian* brings
Hither by Art; Spirits inclos'd attend
Their several Sphears, and with set motions bend
The living work; each year the feigned Sun,
Each month returns the counterfeited Moon,
And viewing now her world, bold industry,
Grows proud, to know the Heavens her subject be,
Believe *Salmonius*, hath false thunders thrown,
For a poor hand is natures Rival grown.

    The reason of this Fabrick, why made of glass Card. *in his Book of Subtilties gives at large. In Philosophy the Doctrine of Reflections, And Refractions, to discover the effects, and affections of air and water, and other liquours, and their various motions, in* Tubes *and* Syphons. *Experiments of a* vacuum *with Mercury, as also infinite experiments of rarefaction and condensation, in* Thermoscopes, *in the* Hydraulicks, *and* Pneumaticks, *in the* Florentine *and* Roman *experiments, and also the* Magdeburgical, *which gave occasion to that rare invention of M.* Boyle, *whereby he hath demonst-rated so many rare conclusions, and tryed so many singular experiments which have made him famous here to all natives as also to all forein Embassadours and learned men abroad.*

Neither may I omit Burning-glasses, nor those for the admission of species into a darkned room, whereby hath been taught the true nature of vision by Plempius *and* Scheiner, *and also by other Glasses the demonstration of the generation of the Rain-bow by* Des-Cartes.

Neither may I forget those Beads, Bracelets, Pendants, and other toys, which have procured us good store of Gold from Guiney, *adorning the Noses, Ears, Lips, Rists and Legs of that nation.*

Glass also affords us Ornaments for our houses and Churches, wherein all natural and artificial things are set out, to the life, in most glorious and Oriental colours.

We shall conclude with that Triangular Glass call'd the fools Paradise, though fit for the wits of wiser men, which representeth so lively Red, Blew, and Green, that no colours can compare with them. And shall relate out of Tagaultius *in what great account the wisest nation accounted the Chineses had of them,* Riccius, *the* Jesuite *fell sick at the City Tanian in China of a most dangerous sickness, But his friend Chiutaiso performed so good offices to him, that within a moneth (which time he stai'd there) he recovered his strength so well that he seem'd stronger than he had been before.*

Riccius *recompensed his friends civilities amongst other presents with a Triangular Glass, wherewith he was much delighted and to add some state to the Glass, he put it in a Silver case, and fastned Gold chains to the buttons at the end of it, writing an elegant Encomium on it, whereby he endeavoured to prove that this Gem was a fragment of that matter whereof the heavens are made. These Ornaments caused many to desire it, for not long after one is said to have offered five hundred Crowns for it. But he then refused to sell it, though he much desired to do so for this is reason onely, because he was not ignorant that such a Glass was a present for the King, and he feared the buyer would prevent* Riccius *by sending it to him, and that the novelty being passed 'twould be less esteemed by that Emperour. But afterwards when he knew that such a gift had been presented to the King,*

Epistle to the Reader. 231

and having somewhat encreased the price, he sold it, and with that sum paid many debts, and thereby obliged his society.

Concerning the malleability of Glass, whereon the Chymists build the possibility of making their Elixir, take their weak foundation from Pliny, lib. 36. cap. 26. They report, *saith he,* that when Tiberius was Emperour there was invented such a temperament of Glass that it became flexible, and that the whole shop of the Artificer was demolished, lest the prices should be abated of the metalls of Brass, Silver and Gold, and this report was more common than certain. *Now* Pliny *liv'd in the time of* Vespatian, *who was the third Emperour from* Tiberius, *so that it appears this report continued long. Many after him relate the same, though with some difference.* Dion Cassius, *lib. 57. thus,* At that time when a very great Portico at *Rome* inclin'd to one side, a certain Architect (whose name is unknown, because Cæsar through envy forbid it to be registred) strangely set it upright, and so firm'd the foundations on every side, that it became immoveable; *Tiberius* having pai'd him banished him the City, but he returning (as a supplicant) to the Prince, wittingly let fall a cup made Glass, and when it was broken remade it with his hands, hoping thereby to obtain pardon; but for this very thing he was commanded to be put to death. Isidorus *affirms that the Emperour in a chafe hurl'd it upon the Pavement, which the Artist took up being batter'd, and folded like a vessel of Brass, he then took a Hammer out of his Bosom and mended the Glass, which being done the Emperour said to the Artist,* doth any one else know this way of making Glass? *when he had denyed it with an oath,* Cæsar *commanded his head to be cut off, lest this being known, Gold should be esteemed as dirt, and the prices of all metalls should be abated. And indeed if vessels of Glass did not break they would be better than Gold or Silver. These three grave Authours,* Pancirollus *and others follow, onely telling it as a hear-say; but* Mathesius, Goclenius, Valensis, Quatriami, Libavius, *and all tribe of the Chymists, assert it with great confidence affirming that it was done by the vertue of the Elixir; but for all this confidence of theirs,* Pliny *onely relates this story with a* ferunt, *they report, and with a* fama, *the report was, and thirdly,* crebrior quam certior, *more common than certain. Which thrice repetition of such like words, sufficiently argue his small belief of the story. It had been enough to have introduced this improbable relation the usual way with a* ferunt, *and hereby sufficiently have provided for his reputation, but he superadds* de proprio, fama crebrior, &c. *which at most proves onely that some small credit was by some few given to it, but* ex vi verborum *a disbelief in the wiser sort. For what can such words as these* (they say such a thing, but the report is most uncertain) *import, but a diffidence in the relator? And 'twas but a* fama, *no Naturalist, Poet, nor Historian deliver it, no record of the person, nor unusual punishment, which is strange, when their Books abound with observations of whatsoever rarely happened. And is it probable that the Emperour himself should not lay up this Glass as a secret in his choicest Archives, and have transmitted it down to his successors, as a thing worth the keeping, being the first of that nature ever made in the world, and perhaps the last, the Artist being put to death? And yet within a few years all this most rare invention, and strange punishment vanish into a report onely. All then was but* vox populi *and* Romani *too, nay, of the cruelty of a* Nero *too, all which might easily keep up this Fable. But why did* Pliny *then relate it! Surely, to please and follow his* genius, *which was to commit to writing whatsoever was rare in Art and Nature, as his nephew in his Epistles, and this Present work witness. Now on this account he might take occasion, in a thing perhaps he judged not impossible to commend that present age (should after times produce any such effect,) and so ascribe the invention thereof to his own nation. Besides 'twas but such a temperament of Glass that rendred it flexible. And is it credible that after ages should not light on't, especially in a thing so commonly practised, and whereto so few but two materials onely are required? Or what means, fame, by the undervaluing of Gold and Silver? I confess I see no inconvenience to the Emperour, nor his Gold and Silvers value, by this invention but many ways advantage, nor any force of consequence in* Cæsar's *words. But so much of* Plinyes *testimony. And what shall the borrowers from him gain more reputation than the first relator gave it? Surely no, especially since they have made such a commentary on* Plinys *text the words will not bear, and have with additionals moulded it into a formal relation.*

Pliny *saith*, ut flexibile esset, *that it might be flexible. Dion comments, the man remade a broken Glass, One degree to malleability, but* Isidorus *completes it saying* with a Hammer he mended it. *Hereby you may see the degrees how this opinion came into the world and by what strange piecings variations and interpretations, it hath been fomented to make that seem credible to after ages, which* Pliny *relates as a vulgar tradition, adding thereto a censure of uncertainty. Which the Chymists to keep up the opinion of their Omnipotent Philosophers stone, omit, and turn* Plinys flexibility *into* malleability. *As if there were no difference betwixt flexible and malleable. Whereas all bodies are in some degree, or other flexible, though none but metalls malleable. A green stick,* Muscovie Glass, *and infinite other things will bow very much, whereon the Hammer, notwithstanding, hath no effect as to dilatation, and formation into thin plates, such as things call'd properly malleable have. And that Glass is in some degree flexible of it's self 'tis apparent, for fine Crystall Glasses made very thin, and well annealed, will bear some small, yet visible bending And I have had* Tubes *made twelve feet long and longer for the* Mercurial experiment, *which being filled therewith would bend exceedingly. So that I am prone to think that if there were any thing at all in this narrative of* Pliny *it might be this, That whereas their Glass before this time was most brittle as being made of Salt-peter, and the art of annealing it (not mentioned by* Pliny*) unknown and consequently must break with the smallest force; Now this Artist might invent and shew such Glass as might accidentally bear a fall, or greater force, than what was formerly made, by making it of* Kali, *and superadding the way of annealing it. which might give occasion to fame, whereof* Virgil, parva metu primo mox sese attollit in auras, *to add some circumstances (which is most common with the vulgar) and so to form this story related by* Pliny.

Now as to the possibility of making Glass malleable, I find not one argument, besides this report, unless by the Chymists *who prove it* per Circulum, *reasoning from their* Elixir *to Glass, and from Glass to the* Elixir. *And surely 'twere more feasible to make the one than the other. For in the making of the* Elixir *the production is* tale ens ex non taliente, *there being no resistance, and incapacity in the matter* ex qua. *But in Glass quite otherwise, for 'tis of it's own nature the most brittle thing in the world, and to make it* malleable *a quality quite contrary to it's nature must be introduced. Besides* diaphaneity *is a property not communicated to any thing* malleable, *and who would call that Glass, that were not transparent? As well may one name that Gold which is not ponder-ous nor malleable, as that Glass which is* malleable *and not transparent. Add hereunto, that the nature of malleability consists in a close and throughout adhæsion of parts to parts, and a capacity to the change of figure in the minutest parts. Both which are inconsistent with the nature of Glass. For the materials of Glass, Sand, and salts, have such figures as seem incapable of such adhæsion in every part one to another. For all salts have their determinate figure which they keep too, in their greatest solutions and actions of the fire upon them, unless a total destruction be wrought upon them, as many instances might evince, and that figure is various according to the salts.* Salt-peter, *and all* Alcalizate-salts *are pointed, and by their pungency, and causticknes seem to be made up of infinite sharp pointed needles, And as for Sand the figure thereof is various, nay, infinite it appears in* Microscopes.

Now how can any man imagine that such variety of figures in Sand can so comply with the determinate figures of salt as to touch one another in minimis *which is necessary to make it* malleable? *Whereas to make it Glass 'tis enough that those two touch one another at certain points onely, whereby such an union it formed, which is necessary to denominate Glass but wholly incompatible with* malleability. *And this union is that which makes in Glass Pores, from whence comes it's* diaphaneity *as you have heard from* Lucret. *Besides something said before, declares that they both remain the same in the compound as they were before. I shall conclude this argument and say, that I conceive that nothing but the* Elixir *will perform this effect, and that both of them will come into the world together.*

Of the Furnaces.

BEfore we enter upon the Art it's self, 'tis necessary to deliver the manner of their *Furnaces*, and their several instruments, and also how their *Metalls* when refined must be wrought, all which are pretermitted by our *Author*, though necessary to be known by the Reader. There are three sorts of *Furnaces* as Agric. *de Re Metallica* distinguisheth them. The first the *Calcar, fornax calcaria* is made in the fashion of an Oven, ten foot long, and seven broad where widest, and two foot deep. On one side thereof, they have a trench about six inches square, the upper part whereof is level with the surface of the *Calcar*, separated onely from it at the mouth by bricks some nine inches wide. Into this trench they put their Sea-coal, the flame whereof passeth into all the parts of this *Furnace*, and reverberated from the roof upon the *Frit*, over whose surface all smoak flieth very black, and goeth out

Of the Furnaces.

of the mouth of the *Calcar*. And the *Conciator* never stirs his *Frit* till the smoak is past. The Coals burn (as in other F*urnaces*) on Iron grates, and the ashes fall thence into the ash-hole, which is level with the floor. The *Conciator* (call'd in the Green-glass houses the *Founder*) is he that weighs, and proportions the Salt, or ashes, and Sand and works them with a strong fire till they run into lumps, and become white, and if the Metall be too hard, and consequently brittle, he adds salt or ashes, and if too soft, sand, still mixing them to a fit temper, which is onely known by the working of it. According as the *Frit* is prepared, he draws it out of the *Calcar*, and when 'tis cold lays it by for use. He doth not here cast water upon the *Frit*, nor water it with Lee, as our *Authour* enjoyns, and after some few days useth this *Frit* to make metall. Which when 'tis melted in the pots, in the working *Furnace*, with a *square*, he rakes and stirs and mixeth well the Metall, when the *square* is red hot, he puts it into a pail of water, for otherwise the Metall will stick to it. With a *Ladle* he takes out the *Sandever*, or empties the Metall from one

Of the Furnaces. 241

pot into another. And with his *Ponteglo* he scums the Glass and with the *Spiei* (an Iron pointed and hooked at the end) he rakes Metall out of the pots for proofs or Essays, to see whether the colour be good, and the Metall fit to work. Some anneal their pots in this furnace as *Agric*.

The second or working *furnace* is that where the pots are set, to which belong the fire place, and ash-hole. This *Furnace* is round, of three yards *Diameter*, and two in height, arched above round about the inside whereof, 8 or more pots are set and on these the piling pots; the number of the pots is always double to the working *Boccas*, that each Master may have one pot refined, and to work out of, and another for Metall to refine in whil'st he works out the pot which hath refined in it; it hath two partitions, the lowermost separateth the pots from the fire place, in the center whereof there's a circular hole made with Iron grates fourteen inches or more in diameter, through which the flame passeth from the fire-place into this *furnace*, from whose arched sides and roof 'tis reverberated into the melting pots; the second partition divides this from the *Leer*, to this furnace belong these holes, first, *Bocca*, the working hole, by which the Metall is taken out of the the great pots, and by which the pots are put into the *furnace*, this is stopt with a cover, made with lute and brick, removable at pleasure, to preserve the work-mens eyes from the vehement heat of the fire; this hath a hole in it more than a palm wide, by which the vessels are scalded as often as 'tis needful. To this *Bocca* belong the *Halsinella's* which are certain hooks, fastened to the sides of the *furnace*, whereon to rest and turn their vessels when they scald them. 2. *Boccarellas*, one on each side of the *Bocca*, lying almost *Horizontally* with it, out of these the *Servitors* take coloured or finer Metall from the piling pot. 3. Ovens or holes next the *Leer* to calcine *Tartar, Iron, &c.* One on each side lying level with the *Bocca*. To this also belong the fire place, having two *Tizzonaios* or stitches one on each side of the furnace, by which a *Servitor* night and day puts in Coals to maintain this *Vestal fire*. These are made with Bricks.

  These *furnaces* are variously made in several places, and to strengthen them are arched with five or more arches, yet all

Of the Furnaces. 243

three are necessary in all *Crystall Glass-houses*. See variety of them, *Agric. de re Metall. l*.10. *Libav. Comment. Alchem. part.* 1. *l.* 1. *c.* 20. *Ferant. Imperat. l.* 12. *c.* 14, 15. *Porta l.* 6. *c.* 3.

The Green Glass furnaces are made square (whereas the former are circular, but where the *Leer* takes off an arch thereof) having at each angle an arch to anneal their Glasses.

The *Leer* (made by *Agricola*, the third *furnace*, to anneal and cool the vessels, made as the second was to melt the Metall, and to keep it in fusion) comprehends two parts, the *tower* and *leer,* The *tower* is that part which lies directly above the melting *furnace* with a partition betwixt them, a foot thick, in the midst whereof, and in the same perpendicular with that of the second *furnace*, there's a round hole (*Imperat.* and *Agric.* make it square and small) through which the flame and heat passeth into the tower; this hole is call'd *Occhio* or *Lumella*, having an iron ring encircling it call'd the *Cavalet* or *Crown*, on the floor or bottom of this *tower* the vessels fashioned by the M^{rs} are set to anneal; it hath 2, *Boccas* or mouths, one opposite to the other, to put the Glasses in as soon as made, taken with a Fork by the *Servitors*, and set on the floor of the *tower*, & after some time these Glasses are put into Iron pans (*Agric.* makes them of clay) call'd *Fraches*, which by degrees are drawn by the *Sarole man* all along the *Leer*, which is five or six yards long that the Glasses may cool *Gradatim*, for when they are drawn to the end of the *Leer* they become cold. This *Leer* is continued to the *tower*, and arched all along about four foot wide and high within. The mouth thereof enters into a room, where the Glasses are taken out and set. This room they call the *Sarosel*, and the *Sarolemen* those who draw the *Fraches* along the *Leer*, and take them thence.

  For green Glass on two opposite sides they work their Metall, and on the two other sides they have their *Calcars*, into which *linnet holes* are made for the fire to come from the *furnace*, to bake and prepare their *Frit*, and also for the discharge of the smoak. But they make fires in the arches, to anneal their vessels, so that they make all their process in one furnace onely.

  The stones wherewith the inside of these

furnaces are made are not brick, (for these would soon melt down into Glass, as also other softer stones) but hard and sandy, which *Imperat.* calls *Pyramachia*, such are brought from *New-castle*, they will strike fire, one being struck against another, and are of a whitish colour. And yet this hard stone frequently rends in a quarter of a year, or else furrows will be made in them. The outside of these *furnaces* are built with brick. The heat of those *furnaces*, is the greatest that ever I felt, and I have observed straws put in three days after the extinction of the fire soon converted into a flame. The workmen say 'tis twice as strong as that in the other *Glass-furnaces*.

  The *melting-pots* come next to be treated of, and are made of clay fetched from *Purbeck* in the Isle of *Wight*, the very same which makes *Tobacco pipes*. This clay being well washed from all impurities is calcin'd in a *furnace* for this purpose, and then ground in their Mill into a fine powder, which being mixed with water is trod with their bare feet till it come to a good consistence, fit to mould, which they do with their hands, and when fashioned, dry them in a convenient place, and afterwards

Of the Furnaces.

anneal them in or over the *furnace*. But those for *Green Glass* are made of *Non-such clay*, mixed with another clay brought from *Worcestershire*, which bears the fire better than that of *Nonsuch*, but both together make the best pots. These pots are fill'd with Metall, and stand level with the Bocca.

Two sorts of pots are used in Crystal furnaces, a greater which will hold three or four hundred weight of Metall, these are an inch thick, and at the bottom, neer two, deep two foot, and above twenty inches broad at the top, but much narrower at the bottom. The second sort of pots they call *piling pots*, because set upon the greater, into which they put their finer or coloured Metall for *rigarines* or other works.

The last business will be to shew the manner of working Glass, which take from *Agric. de Metall. l.* 2. with some additions. The *Servitor* when the Metall is sufficiently refined, puts his hollow Iron into the pot, and turning it about, takes out enough for the vessel or work 'tis intended for, the Metall sticks to the Iron like some glutinous, or clammy juice, much like but more firmly than *Turpentine* or *Treacle*

Of the Furnaces.

taken by tradesmen out of their pots. The figure it takes on the Iron, is roundish, and whil'st 'tis red hot the *Servitor* rouls it too and fro on a Marble that the part thereof may be more firmly united; And then gently blowing into his hollow Iron raiseth the Metall just as blowing doth a bladder or glove. As often as he blows into the Iron (and that must be very often) so often he removes suddenly the Iron from his mouth to his cheek, lest he should draw the flame into his mouth, when he reapplies it to the Iron. Then he takes his Iron and whirls it many times about his head, and so lengthens and cools the Glass, and if it be needful for his designs, moulds in the *stampirons* or flats the bottom by pressing it on the Marble; And then delivers it to the Master workman, who with a gentle force breaks off the collet (which is that part of the Glass which cleaved to the *blowing Iron*, and casts by to make Green Glass) and with his *ponteglo* sticks the Glass and scalds it, and with his *passago* makes the boul of the Glass, then with his *procello* widens and makes it hollower and more capacious, and with the shears cuts off what's superfluous, and

Of the Furnaces.

withall making it plain and even. And thus with *blowing, pressing. scalding* (which must be repeated as often as the Glass cools) *amplifying, cutting,* &c. frames it into the shape preconceived in his mind. And when need requires fastens on feet and handles and with the *Spiei* puts on *Rigarines* and *Marblings*, and when the Master finisheth them another *Servitor* takes them, with an Iron fork and speedily placeth them in the *tower* to anneal, mounting up by a step for the more convenient placing of them, unless by a stumble in the way he chance to break this ware, then most brittle and tender, nay, that will break of it's self without this annealing. So many Masters as there are so many pots at least, and so many *Boccas* or holes there must be, for each man hath his proper station. Where they receive those scorching heats sallying directly into their faces, mouths and lungs, whence they are compelled to work in their shirts like the *Cyclopes* and *nudi membra Pyraemons*, with a straw broad brim'd Hat on their heads to defend their eyes from excedency of heat and light. They sit in wooden large and wide Chairs with two long Elbows, whereon they hang their instruments, fastened so that 'tis immoveable. They work six hours at a time measured by one Glass onely, and then others succeed them, and when these latter have wrought their six hours the former return to their labour, and by this means the *furnaces* are never idle, whil'st they are in good condition, and the pots break not, and the fire keeps the Metall in *fusion*. *Libavius* observes that they are for the most part pale, thirsty, and not very long lived, by reason of their *colliquations*, and the diseases of their head and breast, & that having their bodies weak, they are soon fudled with wine or bear. A very true Character of them.

Observations on the first Book.

Having now dispatched what was necessary to be premised, we come to the text it's self.

    Polverine *or* Rochetta, *are the same thing, and are nothing more than ashes extracted from the same plant, but differing in goodness, as appears by our Author in several places. The name of the latter is wholly unknown to our Glass-houses, and hath now no distinction at* Moran *it's self. The name of* Polverine *still is kept, and 'tis given to all ashes which come from the* Levant *to make Glasses with. The reason then of their different names seems to be that the Polverine was that which was brought in small powder, and the other in hard pieces or stones, and therefore named* Rochetta. *And indeed the workmen observe that the harder and bigger lumps yield a whiter and stronger salt than that which comes over in small pieces or powder. And whether*

Observations on the first Book. 251

this proceeds from the different sorts of this kind of plant, or the seasons of their growth, gathering and burning, or from some sophistication from other salts mixed therewith, or rather from Sea salt water, or other moysture which much endammageth them, I determine not. But certain it is, that to make the strongest salt, and such as will come into hard and stony lumps, they make a Lee of their first burnt ashes, and therewith water the herbs to be next burnt, and so water the herb with new Lees at every burning, and this will make a most strong pot ashes for Soap-boylers, and Dyers. Which way whether it hath been practised to make Rochetta, *and now omitted I cannot affirm.*

Comes from the Levant *and* Syria. Syria *a part of the* Levant. *Now these ashes are brought from* Alexandria *and* Tripoly.

A certain herb. *This herb he names in his Epistle* Kali, *and 'tis so call'd by most Authors but with some small variation, as* Kalli, *and* Kallu, *by* Alpinus, *in* l. de plant. Egypt. *by some* Cali, Alkali *by* Gesner, Soda *by* Lobel, Salicornia *by* Dodon, Salsola, *by* Dodon Gallice, *and* Hist. Lugd. Anthyllis, *by* Camer. Cordus, Fuchs. *and* Lusitanus, *the two latter whereof confidently assert it to be the* Anthyllis *of* Diosc. *both which* Mathiolus

252   Observations on the first Book.

hath fully refuted in his apologie against Lusitanus, *who saw this plant grow at* Tergestum *in* Mauritania, *and lastly,* Antylloides, *by* Thalius. Soda, Salicornia, *and* Salsola *manifestly derive their name* à Sale *from* Salt, *wherewith they all exceed-ingly abound. Of the Antiquity of knowledge, and names of this plant with us, thus our learned Countrey man Dr.* Turner *in his* Herbal. *As I remember it hath no name in* English, *and though it be very plenteous in many places of* England, *yet I could never meet with any man that knew it; But lest this herb should be without a name, it may be called* Salt wurt, *because it is salt in tast, and* Glass-weed, *because the ashes of it serve to make glass with.* Parkinson *saith, that 'tis call'd by the inhabitants of our Sea-coast,* Frog-grass, *and* Crab-grass, *perhaps because those animals feed thereon, being a very juicy, substantial, and not of an unpleasant saltish tast.* Gaspar Bauhin *in his* Pinax *makes ten sorts thereof, whose names and descriptions we omit, as too long for this place. I shall onely mention those three wherewith the* Alexandrians, *and other* Egyptians; *make their* Polverine *for Glass, and Soap, as* Alpinus *chap. 42. delivers them. The first is* Kali geniculatum, *the second sort*

Observations on the first Book.   253

Kali secunda Species, *and* Anthyllis quibusdam, *by* Alpin. *but by* Columna Kali Florid. repens Neapolitanum, *who found it at* Naples, *and figures, and describes it, and saith 'tis used to make Glass. The third sort more peculiar to* Egypt *is call'd by the same Author* Kali Egyptiacum foliis valde longis hirsutis. *And besides these three, I have seen, and have by me a fourth, taken from their Polver-ine bags call'd* Kali spinosum *by the Herbarists. The first and last of these (besides the* minus *and* minimum) our river Thames, *and* Sea-coasts *affords in great plenty, but in no Countrey more* Northerly *than* England, *yet ours will not make ashes for Crystall, or any other sort of Glass, as an experiment made at the Glass-house taught me, for ours being put upon an Iron heated red hot smoaked all away, leaving little or no ashes at all thereon: But the* Kalies *brought from the* Levant, *put on the same Iron, soon converted almost all of them into a very saltish ashes of a dark ash their proper colour, these in burning contracted themselves like worms, flame long, and make a white and very strong salt.* Our Kali *when gathered appears to the tast very brackish and salt, and will being laid in moisture, contract it self into a small dimension, which a*

254   Observations on the first Book.

Confectioner whom I know found to his loss, Who bought thereof instead of Samphire, *For having washed it, and put it to vineger to pickle, found very little of his* <u>Samphire</u> *remaining, for the Vinegar had well near devoured it all. This great difference of plants in respect of the countrey may be manifestly perceived in many other plants as well as in these* Kalies. *As in* Tobaccoes *arising from the same seed, and in* Canary *and* Rhenish *wines from the same stock, in the* Hemlocks *of* Greece, *and those of other countries, and in multitude of other examples, and this reason differenceth* Polverine *and* Barillia. *These* Kalies *though natural denizens of the water, and flourishing naturally neer salt lakes, yet are planted on land in* Spain *and* Egypt, *which doubtless contributes much in those hot regions (especially in* Egypt *where no rain falls, but the Countrey is onely watered once a year, by the rising of the river* Nile, *leaving much fatness and soil behind it) to the quantity, strength, and fixedness of the salt. Now these* Kalies *about midsummer, when in their full strength, are cut down and dryed in the Sun, and then burned, being laid in cocks or piles, either upon the ground, or upon Iron grates, the ashes whereof fall thence into*

Observations on the first Book.   255

a pit where they grow into a hard mass or stone, & are gathered and laid up for use, & are call'd Sode *as* Lobel *affirms. When these plants were first taken notice of is uncertain. The first that took notice of them, and gave them their name, were the* Arabians, *as also to their salt, as appears by their addition* Al *which is purely* Arabick. *Amongst them I find it mentioned by* Serapio *and* Avicen *the Physicians, who both commend it for the Stone, Ulcers, and diseases of the eyes.* Lobel *thinks that we owe the plant, name, and way to make the Salt to the latter* Græcians *or* Arabian *Philosophers Chymists that wrought in Glass.* Advers. pag. 169. *But as to the* Græcians, *and their knowledge of it, I cannot consent, because 'tis not mentioned by any of the* Greek *Physicians or other writers, besides it hath not yet attained any name in that language, and therefore doubtless the* Arabian's *of latter times have conveyed the knowledge thereof to us.*

256   Observations on the first Book.

Chap. 1   TO know the quantities and strength of the Salt. *The best and readiest to know this, is that practised by the Soap-boylers, in their Essay-glasses.* They dissolve their Soap-ashes *in fair water, and* Filtre *the* Lee, *and weigh it again, and so by measuring the quantity of the* Lee, *and comparing it with the weight of the water and ashes before they were dissolved, they find how much Salt such a quantity of ashes contains.*

Brass Coppers. *Our Author every where forbids where the use of* Brass *and* Copper, *unless* Green, *or* Blew *colours are to be made. And certainly these strong* Lees *will fret off some part of the Copper, or else the moisture of the air and Lee will turn part of it into* Verdigreas. *And therefore here they use onely Boylers lined on the inside with Lead, such as the* Alume *and* Copperas *makers employ to extract their Salts in.*

Tartar of red wine Calcin'd. Tartar *call'd by our Author* Greppola *and* Grumi di botti, *which are indeed the* Lees, *and are to be distinguished from the* Tartar *it's self, this sticking to the sides of the vessel in thick hard lumps and (as* Helmont *saith) is never to be found in the region of the Lees, whereas they on the contrary are always found at the bottom of the vessel, moist, and in small pieces onely.* Tartar *of red wine best for this use, because it contains a stronger* Salt, *and more in quantity than that of white wine. 'Tis calcin'd, to burn off all* Hetero-geneous *bodys mixt therewith, and to make the* Salt *whiter, and for the speedier dissolution of it in the water; and better extraction of the* Salt *from the* Polverine, *whose body is opened by the* Tartar, *as* Alume *or* Vitriol *open the body of* Salt-peter, *in making* Aqua-fortis *or* Spirit *of* Nitre, *which otherwise without such like addition would not rise. And for the same cause the* Tartar *must be dissolved in water before the* Polverine *is put into boyler. They* Calcine *their* Tartar *in an oven neer the* Leer, *in the space of six hours, and that to whiteness too, finding that this hath a better effect, than a meaner calcination hath. What advantage the drawing off the humidity of the* Tartar *gives, a secret way used by some* Chymists *doth demonstrate. To make their* Crystals *and* Cream *of* Tartar, *larger, and whiter, they powder it grossly, and then* Calcine, *or rather dry it throughly in an*

258   Observations on the first Book.

Oven, in tin pans. And thus they make them much better, than they can be made without this drying, or moderate Calcination.

The Salt *sinks to the bottom of the boyler and is taken out with a scummer, from which drain all the moysture, and let it run into the boyler; when the faeces of the* Lees *have settled to the bottom of the* tubs, *they draw them this off with a* Siphon.

Chap. 2.   T*Arso. The second material and that which gives consistence and body, and firmness to Glass is sand or stones. As* Iron *gives to* English Copperas, *and* Copper *to* Hungarian, Dantzick, *and* Roman Vitriol, *which otherwise would run into water in moist places and seasons. Concerning these stones,* Agric. l. 10. *saith,* They must be such as will melt, and of them, those which are white, and transparent are best. *Wherefore* Crystals *challenge precedency, For of these broken,* Plin. *saith* Authors affirm, that Glass is made in *India,* so excellently transparent, that no other may be compared with it. The next place, they give to those stones, which though inferior to *Crystall* in hardness, yet are white and transparent, as that is. The third place is given to those which are white, but not transparent. Next to Tarso *our Author commends* Quocoli, *rendred* Pibles, *which* Ferant. Imperatus, l. 24. c. 16. *thus describes,* The Glass stone is like in appearance to white Marble, partaking of transparency, differing from it in hardness, which it hath as much as flint, whence 'tis that being struck, it sparkles and put into the fire, turns not into Lime. *This stone most commonly partakes of a light green, like the* Serpentine stone. *'Tis found in its natural place clad and mixed with veins of crusted* Talk, *when 'tis first put into fire it loseth it's transparency, and becomes white, and lighter, and afterwards it turns into to Glass. 'Tis wrought by the Glass-men, as a material of Glass under the name of* Cuogolo. *Because they gather them in the bottom of rivers, and torrents, in the form of round pibles or shards. And those are they our Author saith are used at* Muran. *'Tis without controversie that all white and transparent stones, such as will not become lime, serve well for Glass but our Authors axiom is not wholly true, for 'either the stones from* New-Castle, *mentioned in the Glass furnaces, nor fire stones, nor rance stones, and many other which strike fire*

260 Observations on the first Book.

with a steel, or horse shoes, and Coaches wheels will not serve to this purpose. Flints indeed have all the properties, and when calcined powdered, and serced into a most impalpable powder, make in-comparable pure, and white crystall Metall. But the great charge in preparing them hath deterred the owners of Glass-houses from farther use of them.

  Sand *is made use of where fit stones cannot be had, and according to our Authors story, were first used; it must be white and small and well washed before used, which is all the preparation of it. Such is usually found in mouths and sides of Rivers; for Crystall requires a fine soft and white sand, but Green Glass, that which is harder, and more gritty. And there is great 'variety in this material, for some soon melts, and mixeth with the ashes and becomes Glass.* Joseph. l. 2. c.9. of the wars of the Jews *relates strange things of* Sand, *which is briefly thus. Neer* Ptolemais *a city of* Galilee, *runs the river* Belus, *arising from mount* Carmel *between* Ptolemais *and* Tyrus. *Neer this small river is the Sepulchre of* Memnon, *having neer it a space almost of 100 cubits most worthy of admiration. For there's a valley round in shew, sending forth*

Observations on the first Book.   261

Sand for Glass, which when many ships coming together have exhausted, the same place is presently filled again. For the winds as it were on purpose, bring from the circumjacent sides of the mountains this Sand. And the place where the metall is, presently changeth into Glass what it hath received. And this seems more strange to me, that the Sands converted into Glass, whatsoever part thereof is thrown into the skirts of that place is again changed into common sand. And Tacit. l. 5. Hist. Belus *runs into the* Jewish Sea, *about whose mouth Sands are gathered* (Lipsius *reads it* Collectae *not* Conjectae) *which having Nitre mixed with them are boild into Glass That shore is small, but unexhaustible by them that fetch it. The same thing witness* Strabo l.12. Plin. l.6. Agric. de foss. *All Authors that write of Glass, mention those places whence the Sand is fetched. Our Glass a houses in London have a very fine white Sand (the very same that's used for Sand-boxes and scouring) from* Maidstone *in* Kent, *and for Green-glasses, a coarser from* Woolwich. *The former will not mix with ordinary green metall. Both these cost but little besides their bringing by water.*

262 Observations on the first Book.

  Cardan l. 5. de variet. *adds Manganese, call'd by him* Syderea, *as a third ingredient into Glass.* Constat *(saith he)* Vitrum ex tribus, *to wit, of stones or sand, of the salt of* Cali, *and Syderea, but the small quantity of* Manganese *added to the metall, can contribute little to a pot of metall. Besides 'tis not used in all sorts of Glass.*

Chap. 3. S*Hews but the common way of Chymists, by solution, filtration, and coagulation to make fixed Salts.*

Chap. 5. M*Ust be cut, &c. All plants have their time to be cut or gathered, that is, when they are in their full growth and strength. The best time is a little before they are in full flour, and that to all purposes, to which the leaves or stalks are used, and also in Chymistry to extract the oyls of Vegetables, and Spirits, which then are produced almost in double quantity more than at other times, but for Salts when the herbs are in seed as* Fern *is at this time. 'Tis a Vulgar error, that* Fern *and other cappillary herbs have no seed, which they have in great quantity on the back-sides of their leaves, in form of dust of a*

Observations on the first Book.   263

brown colour. Nay, Mosses *also abound in seed, as 'tis most evident in an undescribed kind of* Chamapeuce *I keep in my* horto sicco *all whose branches, and betwixt each leaf thereof are multitudes of round and brownish seeds. How much also the seasons of the year difference Vegetables, the* Button-mold-makers *can inform you, in those woods they make molds of, who find that* Pear-tree *cut in Summer works toughest but* Holly *in the Winter,* Box *works hardest about* Easter, *but mellow in the Summer,* Hawthorn *works mellow about October, and* Service *tough in the Summer.*

Chap. 6.   Gives an account of other plants, which yield a Salt fit for Glass In one word, *whatsoever* Vegetables *afford quantity of* Alcalazite *salts (for so the Chymists call such as will persist in, and bear a strong fire, without flying away, and vanishing in the air, and are so denominated from* Alkali, *that is Salt drawn from* Kali), *are good to make Glass. Some whereof this Chapter enumerates.* Kelp *so named from* Kali *and pot ashes are used for crystalline metall.* Kelp *is principally made of that Sea-plant we call* Sea-thongs *or*

Laces, and from it's use by Joan. Bauh. lib.39. c 2. de Hist. plant. Alga angustifolia vitrariorum, *which being kept moist a little after gathering, will shew afterwards, though long kept, it's white salt on the surface of its leaves.* Math. in Diosc. *calls it* Algam vulgarem Venetorum, *the common* Sea-wrack *of the* Venetians, *not onely for the reason before, but also because the Venetians wrapt their Glasses therein which they sent to forein parts. This wrack when the Sea is tempestuous* scopulis illisa refunditur Alga, *as* Virg. *is thrown and scattered upon the Rocks, in great abundance, and also on the shoar, which country people in the summer rake together, dry it as they do hay, by exposing it to the* Sun *and* Wind, *and so turning it as occasion serves till 'tis fit to burn, and make these ashes call'd* Kelp, *used as well to make Alume as Glass. Nor is this particular wrack alone used, though very much abounding in all our Seas, but all other* Algas, fucus & quercus marina *and other Sea. plants, all which abound in Salt.* Pot-ashes *come from* Poland *and* Russia, *and* New-England, *and are the ashes for the most parts of* Firs *and* Pines. *For Green-glasses in* England, *they buy all sorts of ashes confused one with another, of persons who go up and down the*

Countrey to most parts of England *to buy them. But the best and strongest of all* English *ashes, are made of the common way* Thistle, *though all thistles serve well to this purpose, Next to Thistles are Hop-strings, that is, the stem and branches of Hops, cut after the flours are gathered, these two are of late invention.* Bramble-bush *yields the best Salt among trees, and* Genista Spinosa, *and* Hawthorn *next that, and* Kali Spinosum *amongst the Sea-plants. So that it seems that those plants which are thorny and prickly afford in their kind the best and most Salt. Next to the fore-mentioned are all bitter herbs, as* Hops, Worm-wood, Carduus benedictus, Centauries, Gentians, Southernwood, Tansey, Woad, &c. *could store of their ashes be procured at small charges; add to these* Tobacco, *which affords abundance of Salt, the stems being gathered and burnt, and might turn to great profit, though some damage to the soil. A Merchant told me, he offered to* King Charles *the first that he would erect and maintain at his own charges Churches, and endow each thereof with 100 per annum, onely for the stalks of all the* Tobacco *which grew in* Virginia, *and did demonstrate to me the great profit would arise to him by this Patent.*

266   Observations on the first Book.

In the next place follow all Leguminous *plants, such as bear* Peas, Beans, &c. *which have some affinity with the other tribe, especially* Lupins, Fetches, Cicers, *and* Lentils, *the last whereof being lately sown plentifully in* Oxford-shire *for their catel, have been found by experience good to this effect. Add amongst the milky plants, all the sorts of* Tithymals *or* Spurges, *and* Fig-tree, *which have a Blistering faculty in them, and the* Vine-branches, *and* Sow-thistles, *which are somewhat prickly and downy flower'd, wherein they agree with thistles, and have a milky juice, as* Tithymals *have. Now concerning these fixed Salts, observe, that those are best, which are freest from earth, sticks, and all other* Heteregeneous *bodies, and are in the hardest and whitest lumps and to the tast most sharp. Secondly, the best ashes being most full of pure and unmixt Salt soonest run in the* Calcar, *Thirdly, That ashes made with* Vegetables, *when in their full growth, and of the most flourishing branches of them, are best. For from hence the* Chymists *seem to derive their name of* Cineres clavellati, *from* Clavola, *instead of* Clavolati. *Whereof* Varro l. 1. de re rust. c. 40. *thus.* In oleagineris seminibus Videndum, ut sit de tenero ramo ex utraque parte æquabiliter præcisum, quos alii clavolas alii taleas appellant ac faciunt circiter pedales. *Where he expounds* Clavola *by tender branches.* Nonus *reads it* Clavula, *and defines them the cutting of wood. Certain it is that* Clavola *or* Clavula *comes from* Clava *which is our* Club *in* English. Fourthly, these Salts must be kept dry, for moysture, and wet much endamage them. Lastly, some of these ashes make whiter Glass than others. Oak ashes partaking of a Vitriolate nature *make Glass of a darker colour, and* Ash-tree, *and* Hawthorn, *communicating in their Salts with* Niter, *render a more whiter metall than the former.*

Agricola *thus of the Salts* that *make Glass, The first place must be given to* Salt-peter, *the second to white and transparent* Fossil Salt, *the third place to the Salt of the ashes of* Anthyllis *or other Salt herbs, some there are who give precedency to the ashes of* Anthyllis *or* Kali, *not to Salt of Nitre. But those which want make Glass of two parts of the ashes of common* Oak, *or the* Ilex *or* Scarlet-oak, *or* Cerus *the* Bitter-oak, *or for want of them with the ashes of* Beech *or* Fir, *with one part of gravel or sand, and they add a little salt extracted*

268   Observations on the first Book.

from Sea Salt-water, and a little Manganese, *but these ashes make a Glass less white and transparent. Now these ashes are to be made of old trees, whose trunk when grown to six foot high is hollowed, and fire being put into the cavity, the whole tree is burnt down and turn'd to ashes. This is done in the winter when the snows have long continued, or in the Summer when it snoweth not. For rains at other seasons of the year make the ashes foul, by mixing earth with them. Wherefore in the Winter they make ashes of those trees cut into schides and burnt within doors. So far* Agric. *But time and experience have worn out the use of* Salt-peter, *and* Fossil *Salts, which have given the priority to* Polverine, *being too soft and gentle, whereas Glass requires* Lixivial, *and fixed Salts, that have a* Caustical, *and strong tast, and but little* Unctuosity, *whereof* Nitre *and* Fossil *Salt have store, and therefore run most of them into* Sandever, *unto which* Nitre *comes somewhat neer in tast and fattiness. But* Agric. *and other Authors seem to mistake* Pliny, *who puts* Nitre *for those* Alcalizate *Salts, for so* lib. 31. cap. 10. Quercu crematâ nunquam multum nitrum factitatum est, *never much* Nitre *was made of*

Observations on the first Book.   269

Oak *burnt. And* Virgil *also seems to use the word in the same sense,*

I Georg.

Semina vidi equidem multos medicare serentes
Et nitro prius, & nigrâ perfundere amurca.

*I have seen many would anoint their grain
With* Nitre *first, then lees of oil would spread.*

This kind of good Husbandry he expresseth before when he saith,

Arida tantum

Ne saturare fimo pingui pudeat sola; neve
Effetos cinerem immundum jactare per agros.

*Nor with rich dung spare hungry grounds to feed,
And unclean ashes on poor Champains spread.*

As Mr. Ogilby *well renders them. Now these latter verses manifestly prove that salts enrich the soil, and therefore it seems that* Nitre *in the former verses must signifie either salt extracted from ashes, or ashes themselves wherein the salts lye. And to the same purpose are those verses in the same Book.*

270 Observations on the first Book.

To burn dry stuble on the barren fields,
In crackling flames, oft handsome profit yields.

From which burning nothing but salt is produced, whose nature 'tis to destroy the weeds, which being a long time and strongly rooted in the earth, would draw away from the new sowed and tender corn all the nourishment, and thereby render the ground barren, and the seed unprofitable, besides the use of ashes and salt, to destroy worms, which otherwise might eat up the grain. But the coldness of Nitre, *as my Lord* Bacon *affirms, is an enemy to all sorts of grain; Besides learned* Caesalpin. lib.3. cap. 23. de metall. *Calls the ashes of* Kali *a kind of* Nitre. *Add hereunto, that in the* Western *parts of* England, *these Algas whereof* Kalp *or* Kelp *is made to serve the husbandmen to stercorate their land. Which is practised also by the inhabitants of the* Mediterranean, *as* Ferantes Imperatus *relates. And though* Nitre *may be extracted from* Sea-water, *and some* Vegetables, *yet 'twould run almost all of it into* Sandever, *being put into the* Furnaces.

Observations on the first Book.   271

Chap. 7.   SAlt of Lime. *'Tis not here used, that which is sometimes found on old walls, & thence call'd* Paretonium, *is much stronger than the Ordinary salt of lime, a large piece whereof I have amongst my* Cimelia, *very* Diaphanous, *very like in figure to* Alume, *and of a strong* Saline *tast.* Ferant. Imperatus *commends the Lime made of the* Pisces crustacei *and* testacei, *such are* Oysters *and* Crabs *or* Lobsters, *to extract a good salt for glass. And upon experience I have found that a lime of them used in* Holland *by the plaisterers affords plenty of a strong salt: But this salt, though it make a very white glass, yet 'twill not be so transparent as that of* Kali, *and most thereof will run in the pots into Sandever.*

Chap. 8.   FRit, *seems to be derived from* Frittare *to Fry. For 'tis nothing else but salt or ashes fryed or baked together with sand, and so the* English *call the whole quantity baked at a time in the* Calcar *a batch. And secondly the* Frit *melted runs into lumps and like* Fritters *call'd in* Italian Frittelle *or little* Frits. *'Twas by some anciently call'd* Hammonitrum, *and by others more*

272   Observations on the first Book.

agreeable to Etymologie Ammonitrum, *com-pounded of* αμμος, *sand, and* νι'τρον, Nitre. *For so* Pliny, lib. 36. cap. 26. *Fine sand from the* Vulturnian *sea is mixt with the weight or measure of three parts of* Nitre, *and being melted 'tis carried to other Furnaces. There a mass is made which is call'd* Ammonitrum, *and this being reboil'd makes pure and White glass; and* Cæsalp. *more expresly,* Ex arena & nitro fit massa quam Plin. Hammonitrum appellat, hodie Fritta dicitur, *of sand and Nitre a mass is made which* Pliny *calls* Hammonitrum, *but at this day 'tis called* Frit. *This making of* Frit *serves to mix and incorporate the materials well together, and to evaporate whatsoever superfluous* Humidity *they contain in them.* Green-glass Frit *compounded of grosser materials requires 10 or 12 hours baking more or less, according to the goodness, softness of the sand, and ashes.*

    We have three sorts of Frits. First of Crystall *for* Crystall *metall made with salt of* Polverine *and sand. The second and Ordinary* Frit *is made of the bare ashes of Polverine or Barillia without extracting the salt from them, this makes ordinary white or Crystalline metall. Thirdly, Frit for Green-*

Observations on the first Book. 273

glasses, made of common ashes without any preparation of them, or else of Cobbets ground to a fine powder, and a hard sand fetched from Wooll-wich in Kent.

The material must be finely powdered, washed and serced, and then mixed, and equally com-pounded together, and then the fire of the calcar will exactly mix them in the smallest particles and minutest atoms. For otherwise the Sand and Salt; will in the melting pot easily separate the one from the other, which they are apt enough to do were they not often stirred with the Rake.

Pounded in stone Mortars. *This following way now in use is much more expedite, they now grind their ashes which is in hard lumps, their* Manganese, Zaffer, Collets *calcined,* Clay *and* Salts, *in a* Horse-Mill, *the stone whereof is 9 or 10 inches thick, and 7 or 8 foot Diameter, and turns on a floor, where the materials to be ground are put, and are both of hard Marble. This grinding dispatcheth more in one day than 20 men can do in a Mortar.*

We use no casting of water on the Frit *nor wetting it with Lees, but work it off in the pot.*

274   Observations on the first Book.

within few days, if need requires it, though the latter of these two must needs conduce much to the puritie of Glass.

Chap. 9. THe quantity of the Manganese, &c. *the reason whereof is because the colours to be put in are of various goodness, some higher, and others lighter especially the difference of* Manganese *and* Zaffer, *is so great, that some thereof is good for little, other very rich, some of a middle nature, and to know their difference in goodness, there's no way found out but tryall in the furnace, neither in our Glass houses nor in pottery where they have very great use of both. Besides, the metalls of the same materials, and of the same preparation, change the quantity of the colours, in several pots. Wherefore the* Conciator *always puts in all his colours not by weight, nor measure, but by little and little at a time, and then at each time mixeth them well with the metall, and taketh out a proof, and by his eye alone judgeth whether the colour be high enough, and when too low adds more of them till he attain the desired colour.*

The furnace must have dry & strong wood. *Our Author every where commends Oak, for that makes a strong and durable fire with a good flame.* Ferant. Imper. l. 14 c. 16. *saith, that the Glass-men in working-glass because they would have a substantial and gallant rather than a great flame, use the trunks of ash, which ascending directly, and streight, comes to the* Vortex *of the furnace and communicates it's force to the pots within. Ash indeed affords a most pleasant fire but soon decays, and therefore unless a continual supply be made, the metall will not be kept melted, nor fit to work.* Camer. *in* horto *deservedly commends Juniper as a most lasting and strong and sweet fire, could plenty thereof be had. I know not therefore what* Pliny *means, who* l. 36, c. 24. *saith*, levibus vitrum aridisque lignis coquitur. *Glass is boil'd d with light and dry wood. Nor why* Plutarch *should say that* Tamarisk *is fittest to form Glass: for certainly so great a fire as Glass requires cannot be made with such woods. One effect of the fire mentioned by the Arabian physicians, and from them by others, I may not omit* viz. *the burning of Glass mixt with sponge which being g calcin'd they commend to break the stone in the* Kydneys *and* Bladder, *and for outwards* Ulcers. *But the ways to burn it, taught by them, are wholly unfit, and 'tis most certain that the longest and strongest fires, will onely keep it in fusion, but never reduce it to a powder.*

Casteth forth Sandever, sal Alkali, *call'd by the* French, Suin de verre, *that is the fat of Glass, and by contraction in* English Sandever. *'Tis a very white salt, and inclining neerest to a nitrous tast, and easily dissolveth in the air, or any moist place. Our* Conciators *never cast the metall into water, to separate this salt from it, but take it out with a ladle, for it swims on the top of the metall. This must be separated and all scummed off, or else 'twill make the Glass unfit for working, very brittle, and no way plyable. The best metall will yield in a pot of two hundred weight a quarter or half a hundred of* Sandever. *The weaker the salt or ashes are the greater quantity of* Sandever *they yield, some four or five parts more than others do. For green glasses when the ashes are bad they are compell'd to fill the pot four or five times with more fresh ashes, by reason of the quantity of Sandever that is in them, before the pot will be filled with metall. Whil'st any of it remains in the pot, they may not cast in any cold water to hinder the boyling over of the metall, for if they should, the furnace and pots would be blown up together.* Sandever *serveth to make metalls run, and a little thereof put into* Antimony *and* Salt-peter, *for making* Crocus Metallorum, *encreaseth the quantity of the Crocus, and 'twill therewith separate better from the* Scoria. *'Tis sold into* France, *and there used to powder their meat, and to eat, instead of common salt; a solution hereof bestowed on garden-walks destroys both weeds and vermin.*

Necks of the Glass, are also call'd Collets, *which they always break off their iron rods (whereunto they stick) before they take new metall Out of the pot, and these they throw into a place ordained for that purpose, which they grind, and put to the metall, and make thereof the purest green Glass onely, though the product of the finest* Virgin metall.

Chap. 10. Calcine it well. *The Glass must continue twenty four hours or more nay, two or three days in a strong fire the longer the better, for this refines the Glass, and takes away all Blebs and Blisters from it.*

Observations on the first Book.

Chap. 11. T'Artar in great lumps. *Because this comes from the strongest wines, and hath suffered no damage by salt water, or any other, which dissolve it into small pieces, and draw from it some of it's strength. They calcine it in a place made for the purpose neer the* Leer *of the furnace, on either side of the utmost working holes, in six hours time and that to whiteness too, which worketh all the effects in Glass with us, better than a lower calcination doth.*

Chap. 12. ZAffer and Manganese have *no other preparation here than bare grinding them in the mill to a fine powder, and sercing them in the same serces wherewith they serce their* Polverine, *and other materials. What Zaffer is I cannot find in any Author, few there are that mention it.* Cardan l.5. de subtil. *calls it an earth.* Est alia etiam terra quae sic vitrum tingit Caerulei coloris quam Zapheram quidam appellant. *There's another earth which colours Glass Blew, some call it* Zaffer. *But since him* Cæsalpinus, l. 2. c.55. *reckons it among stones.* Alius est lapis vitrum tingens colore cœruleo & si plusculum addatur inficit nigredine, Zafferam vocant. Hic ex cinereo tendit ad purpureum ponderosus & friabilis est; per se non funditur, sed cum vitro fluit aquæ modo. *There's another stone colouring Glass Blew, and too much colours it Black, they call it* Zaffer. *It enclines from an Ash to a Purple colour, 'tis heavy and brittle, it melts not of it's self, but with Glass it runs like water.* Aldrovand. in Museo *follows both, and in one place calls it an earth in another a stone.* Ferant. Imperat. l 26. c. 8. *likens it to the Load-stone and* Manganese. *But 'tis not an earth, for it mixeth not with water, nor will it be compounded with it. Neither is there any stone so brittle as* Zaffer *for with your fingers you may easily crumble it into a sandy gritty substance, which appears so to the teeth. And certainly were it either of these or any natural colour it could not but have been taken notice of by some writer on these subjects, being a thing so commonly used, and and so much thereof spent in Glass and Pottery. It scapt the knowledge of the diligent* Agricola, *who nowhere mentions it and* Jul. Scaliger *who saw a Book concerning Glass, replies nothing to* Cardan *concerning it. So that it seems to me to be an artificial thing of late invention, and made by some metal-men in Germanie (from whence all of it comes) and kept by them as a secret And if I might conjecture at it, I should think that 'twere a composition of* Brass *and Sand, and perhaps some* Lapis Calaminaris *added thereunto. The Blew colour it gives, induceth me to think, that 'tis from brass, as the colour of* Manganese *is from Iron: for certainly nothing can give a tincture to Glass, but what is metalline, and all metalls do give a tincture thereunto.* Lapis Lazuli *a very hard stone loseth it's colour in the fire and so do other pretious stones. 'Tis true,* Antimony *gives Glass a colour, but 'tis by reason of it's Metalline part the* Regulus *onely. Much less will any sort of earths bear the strong heat of their furnaces. For though* Scots-ochre *and* India-red, *may be both calcin'd into good colours for the uses they are employ'd for, yet in the Glass furnaces they wholly lose them. It remains then that nothing but what's metalline must produce this colour, and if metalline what can it else be but* Brass? *For though silver be said to afford this colour, yet that proceeds from the allay of* Copper *wherewith 'tis mixed. For purely thrice refined Silver gives no tincture at all to the parting water. A second ingredient into* Zaffer *is sand your tongue and teeth may easily discover it, but if you put it into* Aqua fortis *you shall manifestly see some white and transparent gravel, very like the powder of our transparent* Pebles, *or perhaps the forementioned* quocoli *described by* Imperatus, *and some other like our common sand, of a brownish colour, which will easily vitrifie. And thirdly, the reason I suppose that* Lapis Calaminaris *may be admixt therewith, is, because neither* Aqua-fortis *nor spirit of* Vitriol, *poured on the* Zaffer, *have any operation sensible thereupon, either as to raising bubles, solution, or tincture. Both which experiments I tryed with ordinary* Aqua-fortis *and spirit of* Vitriol, *and could not perceive the least buble arise, nor smallest motion of these liquours, nor any tincture in either, nor hissing noise, which hapneth in the solution of metalline bodies. But that the* Lapis Calaminaris *hinders the solution & consequences thereof will be manifest by an experiment we shall presently produce. Besides this ebullition may be hindered by the admixture of some Rosin or Gum, on which these liquours have no effect. With what preparation of* Brass *or* Copper, *this is made, I cannot determine, whether from the ore or some preparation delivered by Authors, or what other way, a few experiments might detect this secret, and unty this knot, whereunto I shall leave the Reader. Lastly, whosoever shall*

Observations on the first Book.

consider the weight, value, and colour, now changed from the Purplish of the Authors to a Brown (for so is all that I have seen) will not with much difficulty be perswaded to be of my conceit. 'Tis call'd Zaffer *from the* Saphyre-stone *with whom it communicates in it's Blew colour.*

Chap. 13. M*Anganese (so call'd from it's likeness in colour and weight to the* Magnes *or* Load-stone*) is the most universal material used in Glass, not onely to purge off the natural greenish Blewish colour so call'd by* Virgil 4. Georg.

Eam circum Milesia vellera Nymphæ
Carpebant hyali saturo fucata colore.

Whereon the Commentator,
Vitreo viridi Nymphis apto.
Which is in all Glass and therefore may be call'd the Soap thereof; but also to tinge it, which it doth with a Red, Black, Purple or Murray colour, nay, 'tis the most universal ingredient into all colours as this present work demonstrateth. Concerning it Cæsalp. l. 2. c. 55. *more largely and very well in these words.* Hoc genus Magnetis hodiè vulgo Manganese vocatur,

Observations on the first Book.   283

ab Alberto Magnesia addi soler ad confectionem vitri, quoniam in se liquorem vitri quoque ut magnes ferrum trahere creditur. Lapis est niger, Magnetis similis quo utuntur vitrearii, Si enim modicum ejus vitro admisceatur illud purgat ab alienis coloribus, et clarius reddit, si vero amplius, colore purpureo. Affertur ex Germania, foditur quoque in Italia in montibus Viterbii & alibi. Meminit & Plin. pseudomagnetis. Inquit enim in Cantabria non ille magnes verus caute continuâ sed sparsâ, nescio an vitro fundendo perinde utilis nondum enim expertus est quisquam; ferri inquit inficit aciem ut Magnes. This kind of Load-stone is now call'd Manganese, *by Albertus Magnesia, tis added in the making of Glass because tis thought that it draws unto it self the liquour of Glass as the Load-stone doth Iron.* 'Tis a Blackstone *like the load-stone, the Glass men use it. For if a little thereof be mixed therewith it purgeth it from improper colours, and makes it clearer, but if too much it colours it Purple.* 'Tis brought from Germanie, *'tis also dug in* Italy *in the Mountains of* Viterbium *and elsewhere* Pliny *also mentions the* Pseudo-magnes. *He saith in* Cantabria *not the true Load-stone in a*

284   Observations on the first Book.

continual but scattered rock, I know not whether it be as good to run-glass, for no body yet hath made tryal of it, it colours (saith he) Iron as the Load-stone doth. Cardan. l. 5. de Subtilitat. *calls it* Syderea *(upon what ground I know not) and mistakes the colour, putting Blew for Red. whereunto* Scal. exerc. 104. 23. *replies,* Manganese *is unknown to me, yet in a Manuscript of blowing Glass belonging to* Pantheus *a* Venetian *'twas written, that Glass was coloured Purple therewith. Believe the Author as you please. I remember when I was a Boy and lived at* Ladroni, *there was dug up at the* Solodonian-mountains *(if I mistake not) I know not what, which they said was carried to* Venice, *wherewith Glass was refined to that whiteness, and purity that it kept the name of* Crystalline. *I seem to remember the colour was that of Iron.* Secundus *my Master taught me that Glass by the admixture of an Iron colour grew white by reason of the strange* Cohæsion *of both substances, whose parts being compounded, the colours also entred one into another and that the* Manganese *of an Iron nature did exhale, being impatient of the fire and carried away with it the foulness of the Glass, no otherwise than Lees wherewith linnen is cleansed. A judgement not unlike this opinion*

I find in Arist. *where he sheweth the force of* Origanum *to purge wine. But this Iron substance exhales not, if it be mixed with metalls, because then 'tis baked with less fire or a less time. And this is all we have delivered concerning this* Manganese. *Now in these discourses, two things are observable the attraction, and purgation. As for the former, attraction of the liquour of Glass, there's no ground for it, no more than the bare name imports, which was imposed* ex placito: *For if you apply never so great a quantity of* Manganese *to the smallest particle of broken or melted Glass, it stirs it not. And then if they mean by the liquour of Glass the* Sandever *part thereof, 'tis certain the greenish colour remains in the metall after that is wholly scummed off, and that Manganese then put in refines it. But if they mean by liquour of Glass onely liquid Glass, then 'tis onely* gratis dictum, *no argument, no experiment being brought to prove it. As for that of purifying 'tis as manifest as the attraction is obscure. Though the* modus *be very doubtful.* Scaliger *and his Master* Secundus *think 'tis by the way of exhalation, and perhaps,* Plin. *&* Cæsalp. *mean by their attraction, this purgation, but then they tell us not what becomes of them both. They must be separated from the metall by*

286   Observations on the first Book.

precipitation or exhalation, but the former cannot be, for then the metall being stir'd twould return to it's former Colour, or 'twould be found in the bottom of the pot in the form of powder, as in other precipitations 'tis constantly usual. And the exhalation is as incredible since there appears to be no loss of weight after this refining, besides, how can the fixed bodies of Manganese *arise in exhalation being inviscated with the tenacious substance of Glass? and what strange choice can there be supposed in the* Manganese, *that it should call out the Greener part onely of the metall to be carried away with it into the air, and in insensible vapors too? The reason seems to me to be onely a change in the figure and minutest parts of the metall, for the fire making the* Manganese *run, mixeth it with the smallest atoms of the metall throughout, which by boyling, and various agitation and revolution of them frames those* atomical figures *which are apt to reflect most of the light which falls upon it, and is the same we call White. Multitude of instances might be given to illustrate this doctrine of the production of colours by mere transmutation of parts, but we shall content our selves with those onely which by admixture of co-operate bodies become White. Take then* Terebinthine *which is of a yellowish*

Observations on the first Book. 287

colour, or Oleum Capevæ *of a blackish colour, or tinge oil of* Turpentine *with Verdegreas (in which 'twill easily dissolve) into as full a Green as the natural colour of Glass and shake either of these very well together, with the yolks of Eggs, and they all make a very clear and white colour. Or else take a strong* lixivium *of the* Soap-boilers *and mix it by agitation with the greenish oyl of Elder, and you shall therewith make that medicine Physicians call* Lac. Virginis, *you may do the same with any other oyl, and the said Lee. Here you have the colour of a Yellowish Red-lee destroy the green of the Oyl. Again Oyl of* Tartar *poured on the green water made with the solution of the* Pyrites *in rain water, gives a white colour, nay the said Oyl poured on Green or Blew* Copperas *as dissolved in common water, effects the like, though the Colour will not be altogether so White as in the former, unless you add a great quantity of oyl of* Tartar. *Which instances sufficiently refute the way of exhalation, and manifestly convince that this purging of Glass, is wrought onely by a various texture, and position of the parts of the metall, made by this new accession of* Manganese. *Nay, what other reason can be assigned, but this change, why Salt and Sand both most white, should produce*

288   Observations on the first Book

a coloured metal? or why Zaffer *and* Manganese *should produce a Black?*

That Manganese *consists of much Iron seems beyond contradiction, which may be evinced by these experiments. I poured* Aqua-fortis *upon some powder of it, and in a narrow mouth'd glass, the water rose up in great bubles, and immediately boiled over the Glass, and in a Glass, with a wider mouth it rose less, and a strong, and most piercing fume there from, offended much my Nose-thrils. And Spirit of* Vitriol *poured on it boild, a little, but sparkled more, the glass became so hot that I could not hold it in my hand, and that which seems peculiar to the* Manganese, *fair water poured thereon encreased the decaying heat very much. The tincture of this stone was of a deep claret colour. All which agree throughout, with the same Spirits poured on Iron; and certainly the colours of the* Manganese, *come from the iron that it contains. Red is common to them both, and a Purple is but a deeper Red with an eye of Blew, and the same colour some preparations of* Crocus Martis *have, and as black is made with* Zaffer *and* Manganese, *so rich Blacks in silks are made of slip, that is the powder which the Shear-grinders grind from shears and other edge tools mixed with Sand from the Grind-stone, and doubtless would be of use in the colour pots of the furnace did they know it, and would they use it. Secondly, this* Manganese *makes the metall rise much and boil as all Iron or Steel alone, or* Crocus Martis, *or any other preparations, or composition thereof; which quality is also common to* Copper, Brass *and* Lead. *Observe here, that wheresoever any of these are put into the pot our Author commands that it be done leasurely and by little and little, and that some vacuity be left in the pot, for fear you lose your metall which will run into the fire and ashes; and thereby you lose the time and charge, for all this commonly goes together with him.*

Our Author here commends Manganese *of* Piedmont, *for the best in the world, and therefore wherever he mentions the one, he subjoyns the other. But some few years since, the industry of our nation hath found in our own countrey at* Mendip-hills *(famous for Lead) in* Somerset-shire, *as good as any used at* Moran. *Wherever the Lead-Ore-Men find it, they certainly conclude that Lead-Ore lies under it. They call it Pottern-Ore, because the Potters spend such great quantities of it, this being the onely materiall wherewith they colour their ware Black, as they do Blew with Zaffer. They*

290   Observations on the first Book.

count that the best, which hath no glittering sparkles in it, and is of a Blackish colour, but powdered of a dark Lead colour, 'tis very hard ponderous, the deeper the colour, the deeper it colours the metal in the Furnace, 'tis to be put into the melting pot together with the, Fritt.

Chap. 14, 15.   FErretto of Spain, *commonly call'd* aes ustum *or burnt Brass, and 'tis made* Latin, *by* Cæsalp. l.3. c.5. *where he thus saith,* Optimum aes ustum conficiebatur in Ægypti Memphide deinde in Cypro, cujus notæ sunt, ut sit rubrum & attritu colorem Cinnabaris imitetur, nam nigrum, plusquam decet exustum est. Hodie in Hispania conficitur, appellant autem Ferrettum, sed nigrum est, inficit ingredine, ideò utuntur ad capillum denigrandum. *The best burnt Brass was made at Memphis in Egypt, afterwards in* Cyprus, *the marks whereof are that it be Red, and that by bruising it imitate the colour of* Cinnaber, *for that which is black is too much burnt. 'Tis now made in* Spain, *they call it* Ferrettum, *but 'tis Black and colours Black, therefore they use to colour therewith their hairs Black. But if it be calcin'd to a mediocrity it appears Red, & 'tis of the same colour when powdered, and hence it seems to have it's name* Ferrettum à ferreo colore, *for* Crocus martis *appeareth to the eye Red, though much lighter than* Ferretto *doth. By the former <u>discourse</u> of* Cæsalp. *that some Countries afforded better* Ferretto *than others, as* Castile *soap, and* Venice *Glasses are the best, but we find no such difference in the several climates, that we need fetch any thereof from* Spain.

The two most eminent and singular colours, both in them selves, and in relation to animals, and to this Art of Glass, are Blew and Green, in themselves, as partaking much of light, as it seen in the Triangular-Glasses, and they are also most delightful and agreeable to the sight, and eyes of animals as neither widening nor contracting the Pupil too much, both which are dolorous and, offensive; and in the Art of Glass, in Pasts, Enamels, Glass of Lead by reason of their great conformity and neerness to many sort of gems, challenge a great share of use, besides the many gradations of them used simply of themselves, or else blended and mixed one with the other, Blew is a simple colour in all Arts conversant about it, but Green in the curious Art of dying is a compounded colour

292   Observations on the first Book.

of Blew the Ground, and Yellow super-induced, or contrary-wise wrought. But in other Arts this colour is simple, and both arise from the same materiall Copper *or* Brass *by various ordering and preparing them. 'Tis a strange and great mystery to see how small and undiscernable a nicety (though the same materialls be used) makes the one and the other colour, as is daily discovered by the refiners in making their* Verditers *who sometimes with the same materials, and quantities of them for their* Aqua-fortis, *and with the same* Copper-Plates, *and Whiting make a very fair Blew* Verditer, *otherwiles a fairer or more dirty green. Whereof they can assign no reason, nor can they hit on a certain rule to make constantly their* Verditer *of a fair Blew, to their great disprofit, the Blew being of manifold greater value than the Green. Now although the genuine and natural colour of* Brass *and* Copper, *is the true Sea-green, mixed of both colours, yet the former inclines more to a Blew than the latter, and the dissolvents have a great share in this business. For* Verdigreas *made of Copper-plates buried in the earth with Grapes, makes a Green, but* Copperas *made with Copper, and the liquour of the* Pyrites *dissolved with rain water, yields a*

Blew in Dantzick *and* Hungarian, *and* Roman Vitriol, *the onely difference of these proceeding from the resolution of the materials into finer & minuter Particles, and various texture of the* Atomical *parts of the materials dissolved. Now the reason why* Brass *makes a better Blew than* Copper, *seems to be this that the* Lapis Calaminaris *the onely thing that differenceth them, takes in, and incorporates with it's self that acidity which naturally* Copper *contains, which as it appears in the making of* Verdigreas *turns it to a Green, being exalted by the acidity of the Grapes. And this seems also to be the cause, why* French-wine-grapes, *which have more acidity in them than* Spanish-wine-grapes *have (though the climate of* Spain *be more suitable than that of* France*) are fittest to work this effect. The force also of* Vitriolate *juyces may be seen in our* English Copperas, *and* Vitriol *of* Mars, *made of* Spirit *of* Vitriol *and* Steel, *both which change the natural Yellowish colour of Iron into a Green, and* Lapis Armenus *a Blew stone ground with Vineger, or the tincture thereof drawn. The effect of* Calaminaris *in drinking in the acidity of the* Vitriol *do the same, an ingredient into* Aqua fortis *is clearly manifested, by a pretty and lucid experiment,* *was once shewed me by my neighbour a Refiner, who bought some* Copper-plates *to draw down his silver from* Aqua-fortis *wherein 'twas dissolved, but these* Copper-plates *would not wholy præcipitate the said silver, but left ten pound thereof in thirty remaining unpræcipitated in the water. The reason whereof was found to be, because the* Copper *for those plates had been melted in a pot, wherein* Brass *before had suffered* Fusion. *The* Copper-smith, *hereupon remelted the said plates in new pots, and with a strong fire, burnt off (as they usually do) the flours of the* Lapis Calaminaris *which are volatil and fly about the work-house, colouring the Cloaths, hairs, and Beards of the Work-men, as white as those of Meal-men, or Millers. Now when these flours had been well separated, and the* Copper-plates *freed totally from them, they drew down the silver wholly from the* Aqua-fortis. *Now in this experiment the* Lapis Calaminaris, *imbibed part of the acidity from the* Copperas, *and so the Plates being less corroded, and consequently too little thereof received into the parting water, left room for the silver to remain, and to be supported by the said water which is the reason of all præcipitation, for the new advenient metall coming into the place of*

Observations on the first Book.   295

the silver, forceth it to descend upon the Boule and Plates in the form of a white powder. But that this effect followed from the imbibition of the acidity from the Aqua-fortis *seems manifest, because* Aqua-fortis-vineger, *or it's* Spirite, *or any other acid Juyce, poured thereon becomes more sweet, and heavy, as they do with Coral, Crabs-eyes, (as they are falsly call'd) the shells of fishes or* Lapis Lyncis, *and whiting wherewith & the water from the* Copper-plates Verditer *is made likewise do. And hence it proceeded too that the water made with these Plates, acquired the most singular Sky-colour the said Refiner had ever seen. And to this purpose I remember, that from Brass dissolved in common* Aqua-fortis, *with an addition of* Crabs-eyes, *a most fair Sky colour proceeded thence.*

Of all metalls Copper *is the most plyable to the Hammer, drawing into wire, gives malleability to silver and gold in coins, and is of no hard solution in the fire, is soon corroded with any acid Spirits or Salts, and without great difficulty is resolved into a powder with the fire. Five preparations or reduction to powder our Author gives, First, a calcination of Copper, c. 14. of* Brass, *c. 21. with* Sulphur,

296   Observations on the first Book.

then with Vitriol, *c.* 15. *Thirdly, a simple calcination of* Brass *by fire, c.* 20. *of scales of* Brass, *c.* 24. *Fourthly, scales thrice calcin'd, c.* 25, 28. *Fifthly, the making of* Vitriol *of* Venus, *c.* 31, 132, 133. *All which are so well known to the meanest* Chymists *I shall need to say little of them, especially having given so large an account, how the two prime colours, Blew, and Green are thence educed. But above all these preparations, that of* Vitriol *of* Copper *carries the preheminence, and next to that being prepared the same way with it, the calcination with* Sulphur, *and especially with* Sulphur vivum *in a clear and strong fire makes a better colour than any of the other calcinations mentioned by our Author. For though Originally* Brimstone *and* Copperas *are made of the same* Marcasite, *and produce* Spirits *undistinguishable each from other, yet* Sulphur *sooner and better penetrateth into the body of the metall, being more vehemently driven in by the most acute and sharp points of the flame, and so consequently divide more subtilly the smallest particles thereof. Besides the flame dissipateth and carries off the Spirit of the* Sulphur, *which of it's own nature is apt to blacken, and make all colours more dirty. For as 'tis well known* Copperas *with gal[l]s or any other astringent* vegetable *make Ink, and the Black for dyers. But if you list to try* Vitriol, *you must not use* English Copperas *made with Iron, but that which is made with* Copper, *Because experience teacheth the Refiners that* Aqua-fortis, *made with it will carry 'its foulness through all their mediate solutions even to the* Verditer *'its self, which 'twill make infallibly of a dirty green colour. Wherefore they make their* Aqua-fortis *of* Dantzick Copperas *onely.*

Whosoever then would extract a good colour with Aqua-fortis *(which way our Author useth not though he doth in making* Crocus-Martis*) should make it with* Salt-peter *and* Alume *instead of* Vitriol *as 'tis hereafter made for* Calcidonies, chap. 38. *or with* Hungarian *or* Roman Vitriol *especially the last which makes the strongest water, being most impregnated with* Copper, *and coming neerest to* Vitriol of Venus, *for with these waters rise some small atoms of* Copper *(as 'tis manifest by holding a knife over the fumes of such* Aqua-fortis *boyling) which will colour it of a perfect* Copper *colour. And if you dissolve in this* Aqua-fortis *the best* Copper, *and then*

Observations on the first Book.

precipitate it with speltar (which I have sometimes done with the refiners double water impregnated with Copper*) you shall have a most excellent Blew, which may be of good use for the colouring of Glass, for I doubt not but the strong fire of the furnaces will wholly dissipate the speltar being of a Sulphurious nature or convert it to Glass, for upon the dissolution thereof with* Aqua-fortis *it shooteth into Green Crystalls, however the* Copper *will remain to give it's tincture to the Glass, and that this way of precipitation is much better then by drawing of the Spirit with heat 'tis apparent by this, that the finer and purer parts of the* Copper *rise with the water as in the experiment of the Knife, and by many others to be met with in the writings of the* Chymists. *One experiment more I shall add to extract the tincture from* Copper. *I took* Copper *calcin'd and* Verdigreas *of each an ounce and flll'd two Glass bottles with the juyce and leaves of garden* Scurvigrass, *which abounds in volatile salt, and closed these Glasses well, and first for a month, set them in a Sellar, and afterwards upon Leads [the roof] in the Sun, during the Summer, moneths then I strained the* liquor per chartam emporeti eam, *and had from the former, a fair Skie, from the latter a pure Sea Green.*

And this I the rather relate, because I have not met with any experiment in this nature with volatile Salts, and 'tis very probable that other plants full of the same Salt, especially having some clammy juyce in them, such as Onions, Garlick, Leeks, and Molyes have, might shew some rare effect upon Copper, *for their leaves have either a deep Green, or else a Green mixt with Blew. The whole tribe of* Acids *also are dissolvents of* Copper, *and all sorts of fixed salt, all which have acidity in them. And no doubt great variety might be met withall in diversity of menstruums, and processes of extracting these tinctures.*

Our Author c. 20. tells you Brass is made of Copper *and* Lapis Calaminaris, *I shall here deliver the process since I find it no where fully delivered,* Lapis Calaminaris *is found in* Sommersetshire *and the North of* Wales, *and though some of it hath been brought from* Dantzick, *yet 'tis not of the same goodness with ours of* England. *This stone before* use *must have the following preparation. It must be first calcin'd in a furnace like the* Calcar *with a small hole on one side to put fire in, which may be either of Coal or Wood, but Wood is best, because it maketh the greatest flame, and consequently the best reverberation. The time of Calcination is about five hours, in which space they often rake it about with a great Iron rake, It requireth good judgement to calcine it well: for when 'tis not sufficiently calcin'd 'twill not mix with the* Copper, *and when too much, 'twill make it too brittle, and in both cases gives not the true tincture to* Copper. *The sign of its just calcination is, when 'tis in a white and very fine powder. Almost half of the* Calamie *(as the workmen call it) is wasted and flies away in flour, which sticks to the mouth of the Furnace of divers colours of little use with them, though I could easily prove these flours to be the true pompholix of the ancients, and to be used in the ointment, that hath it's denomination thence. 'Tis an excellent dryer, and applyed to Gleeting Nerves, and Tendons, without pain, it soon <u>exsiccateth</u> them. This powder I communicated to the eternal glory of our nation, and* Anatomy, *& an excellent Chirurgian, and never to be by me forgotten the incomparable Dr.* Harvey, *a man most curious in all natural things, who confessed he thought this to be the said* Pompholix, *and with most happy success frequently used it. Now when the Calamie is well calcind, they grind and serce it to a very fine powder, and therewith mix well*

Observations on the first Book. 301

Charcoal *finely ground to a powder, this mixture they put into the bottom of a pot, and upon it a* Copper-plate, *to wit, seven pound of this mixture, to five pound of* Copper, *which is their usual proportion. These pots are made of* Nonsuch-clay, *which must be first calcin'd if they make pots of it alone, but usually they grind their broken pots with an equall quantity of the clay, and therewith make them, which being well wrought and annealed, will commonly last 12 or 14 days.*

The furnace wherein they melt their Copper *and* Calamie *is about six or seven foot deep under ground, the earth being circularly raised by degrees from the plain of the work-house to the hole, whereby the materials and fire are put into the furnace, which is the center of the raised earth, and in a perpendicular to the bottom, and area of the furnace. The diameter whereof at the bottom is three or four foot wide, growing gradually narrower and narrower in the form of a cone to the said hole which contains a foot in diameter, wide enough to put in and take out their pots and fire. This hole hath an Iron cover with a small hole in it, wherewith they regulate their fires. At the bottom of this furnace, they have a long pipe or hollow*

302   Observations on the first Book.

place by which they blow their fire with bellows. At first they make a very gentle fire, encreasing it by degrees, till they see the Copper *melted down, and well mixed with the* Calamie, *which is usually done in the space of twelve hours, for every twelve hours, they cast their plates at five in the morning and evening; and then they take their pots out of the furnace (which are usually eight or ten in number) with a long pair of tongues, and set them in a hot place a little time till the metal grows a little cooler, yet still melted, and then pour it out of all the pots together into a mould of stone, which produceth a plate of Brass three foot long, and a foot and a half wide weighing betwixt 60 and 80 pound. The mould is made of two stones which seem to me to be of that sort, which are call'd* Calcarii, *for they have many small shining particles in them like Spars, which continue after long use of these stones, whose colour is thereby changed from a Gray to a reddish* Copper *colour, onely the spots remaining. These stones have formerly been brought from* Holland *but have been sometimes since found in the mountanous parts of* Cornwall, *and are as big as a reasonable gravestone, and of the same figure. They must be annealed some hours before they cast their plates on them, else the*

Observations on the first Book. 303

metall will fly, and besides endanger their breaking. They must have many pair of them in readiness, because after three days casting they become weary (as they call it) and must be new coated with coal and tallow. 'Tis to be observed that the mixture of the Calamie *and Coal, must be always put under the* Copper-plates, *for then the Calamie being raised by the mixt* Charcoal *and heat of the furnace easily penetrateth and mixeth by little and little with the* Copper *melted, and so both unite into one mass, making the compound call'd* Brass. *Whereas the* Calamie *would most of it fly away should it be put above the* Copper-plates. *And though the interposition of the Copper hinder its ascent, yet much thereof flyes away and sticks to the sides of the furnace, and according to the diversity of the superior or inferior part of the furnace where 'tis found and difference in figure and colour receives various names, of* Capnitis, Botrytis, Placitis, Onychitis, Ostracitis, *so call'd by* Plin. l. 34. c. 10. *All which contain some* Copper *in them easily discoverable by the affusion of* Aqua-fortis *on them or by long lying exposed to the open air, nay, you shall see in them sometimes a Greenish Blewish colour, when they are taken out of the furnace. The encrease of*

304   Observations on the first Book.

weight by the Calamie *is from 38 to 40 pound in the hundred, so that 60 pound of* Copper *makes with* Calamie *100 pound of* Brass. *Observe also that the fire must not be too strong,* <u>nor</u> *must the pots continue too long in the furnace after fusion of the* Copper *lest the* Calamie *fly away, and that the coals lying at the bottom of the pot, and which were mixed with the* Calamie *are not totally turned to ashes, but oftentimes come out untouched, and unaltered though the pots have continued red hot for many hours together, which is needful because* Copper *with the* Calamie *require longer time to be melted than* Copper *alone doth. As to the easie parting of the* Calamie *from the metall, we shall to what hath been formerly said, add this, that when they draw this* Brass *into* Wire, *at each new drawing they must anneal it else 'twill break, and yet they must not heat it to above a* Cherry *red, for if they do they burn* off *the* Calamie *to their great loss, which is easily done in* Brass *drawn into small threads.*

Chap. 16, 17, 18, 19.   D*Eliver several Wayes of making* Crocus Martis, *All which and many more are delivered by* Chymical *Authors. They may be reduced to* *these heads,* 1. *A simple reverberation without admixture, and such I have seen made of Iron Bars wherewith some furnaces are supported and built, and the best, and deepest colour, I ever saw was made this way in a furnace wherein* Aqua-fortis *was constantly distill'd the whole bars turning by little and little into this* Crocus, *and was brushed off in a considerable quantity. The second way is a calcination or reverberation with Brimstone, Salt, Urine, Vineger. Thirdly, by solution in* Aqua-fortis, Aqua-regis, *Spirit of Salt and* Nitre, *and then by exhaling the waters you shall have a very Red powder. The solution of Iron in Spirit of* Vitriol, *or of* Sulphur *make the* Vitriolum Martis, *not much differing from our* English Copperas *in goodness but onely in strength, either as to dying, or Medicines, which being calcin'd makes a* Colcothar, *not unlike that of common* Vitriol, *which though it may serve Painters for a deceitful colour, yet 'twill not serve the Glass-furnaces, for all* Colcothar *contains in it much terristriety which would make the Glass foul and obscure, this seems to be the reason, why our Author useth not* Vitriol *here, as he doth before with* Copper.

306 Observations on the first Book.

I shall say no more concerning the tincture of Mars, *but that whatsoever of* Acid *or biting juyces work upon* Copper, *the same have also their effect upon it. And though all the ways produce a red, yet some of those reds are lighter and more transparent, than others, and so may serve for several colours, and various admixtures with other* Metalline *colours, to advance or moderate them, for* Crocus Martis *made with Vineger complies with Greens,* chap. 32, 34, 35. *and in the* Emerald *colour of Glass of Lead,* chap. 65. *and for the same colour in pasts 'tis used indifferently with* Verdigreas, chap. 77, 78, 79. *and in Blacks,* chap. 101. *but for a fair red,* Crocus Martis *made with* Sulphur, chap. 128. *but for more fair colours* Crocus Martis *made with* Aqua-fortis, chap. 43. *But so as the best colour from* Brass *is of* Vitriol *of* Venus *the primest and lightest colour from Iron or Steel is that which is made with* Aqua-regis *which proceeds partly from the mixture of* sal Armoniac, *and partly from a finer solution of it.*

And thus having past over the prime materials, and preparations for colours in Glass,

Observations on the first Book. 307

the rest of the work consisting principally in the due mixture of the said colours with the circum-stances, which our author hath fully done, we shall be very brief in what follows, and shall onely deliver here one preparation come to my knowledge, whilst a secret of great value, but now commonly enough known to the furnaces, and 'tis this.

Take of Antimony *and* Salt-peter *well ground and mixed, of each twelve pound, together with 200 weight of the common materials for glass wherewith this mixture of* <u>Antimony</u> *and* Peter *must be also well united, and then calcin'd in the calcar and made into a* Frit, *or which is all one make* Regulus *of* Antimony *with* Crude Antimony *and* Peter, *the manner every* Chymist *knows, which being mixed with the metall afford a very white* Enamel, *and serves with other mixtures for various colours.*

Chap. 29. Port. l. 6. c. 5. *To colour the Blew Gemm which the common people call* Aqua marina *(and our* Jewellers Egmarine*) a kind of* Saphire. Beat burnt Brass *into a most fine and impalpable powder, otherwise a courser gemm will be made thereof, and let it be mixed with Glass. The quantity cannot be determined, for they are*

308   Observations on the first Book.

made deeper or lighter, for one pound of metall one drachm of burnt Brass *will suffice.*

Chap. 32.   FOr the Emerald *colour* Porta, l. 6. c. 5 *thus, when you have coloured that* Egmarine *you shall easily turn it to an* Emerald, *by adding half* Crocus Martis *to the calcin'd* Brass *to wit, if at first we put in a fourth part of* Brass, *we now add an eighth part of* Crocus, *and as much calcin'd* Brass. *Observe that they boil together six hours after the colours are put in the stuff, that the jewels may grow clear which became cloudy by putting in the colours.* Brass *is heavy, and when 'tis mixed with the metal, every moment 'twill sink to the bottom of the pot, and make the gemm more dilute, wherefore you must very often stir it. Let the fire decrease by little and little till the furnace grow cold, let the pots be taken out of the furnace, and being broken they afford you counterfeit jewels.*

Observations on the Author.

R*Osichiero, whereof thus, Port lib. 6. cap. 9.* But the more skilful and modern Glass-men in colouring Enamels of a clear Rose-colour (the common people call it *Rosachiero*) take not a little pains, seeing our Ancestors made it Artificially and beautifully.

Chap. 37. O*Ur author adviseth you to make your* Aqua-fortis, &c. *your self; and good reason for it, for one pound of common* Aqua fortis *upon my reiterated experience shall yield but four ounces of good Spirit, the other 12 ounces will be phlegm of* Vitriol. *This addition of white* Arsnick *in the making of* Aqua-fortis, *I find in the Lady* Isabella Cortese *printed at* Venice *in Italian 18 years before the publication of this work. Many are the compositions of this water, but* Nitre *is the principal operative ingredient* *in them all. Most make it of* Vitriol, *some of* English Copperas *which serves for common uses, and for* Bowdies *(though made of* Dantzick Copperas *would be better for that use) for this the refiners use finding a dirtiness from our* English, *some add alum instead of* Vitriol, *but that yields at best but a weak phlegm, Others have made essays with* Sal gemm, *but they found that this Salt affords no Spirit, but sticking to the neck of the retort, hinders the passage of the Spirits and breaks the vessels. When the red fumes are past all the Spirits of* Nitre *are raised, and then the fire is to be extinguished, for after followeth onely the Spirit of* Vitriol, *which hindereth the operation of the Spirit of* Nitre *towards the solution of metals. I have often seen singular good parting water drawn by the refiners twice in 24 hours in which time with their fire, not much of the Spirit of* Vitriol *could arise, which requireth commonly three days with the strongest fire can be made for the last two days to draw off both the Spirit, and ponderous oyl from it, though the volatility of the* Nitre *in* Aqua-fortis *may help to raise them. One thing our Author omitteth though very necessary to be done before the* Aqua-fortis *be used, practised constantly by the*

Observations on the Author. 311

refiners, else their waters will be foul, the manner is thus, as you have it in Beg. Tyrocin. Chym. c. 3. *Take of the distil'd water and put into it a peny weight of refined silver, and dissolve it upon coals, then pour this silver water into three parts more of the unfined* Aqua-Fortis *which will become of a milky colour, then they let it settle, and decant off the clear, this setling the refiners call the fixes, and cast it into a tub of water of 20 gallons, all which it will in a moment turn to a milk colour. I know a refiner who destilleth his* Aqua-fortis *out of an Iron pot, which he finds to make a stronger water, besides the great charge in pots and fire saved, you may see the way in the commentator on* Beguin.

Chap. 40 *YOu need not charge your* Aqua-Regis *with so much* Sal Armoniac *as it will dissolve, one ounce and a half to a pint is sufficient. I wonder at* Beguins *way of making this water, who distills the* Salt-peter *and* Sal Armoniac *together, but experience hath taught me that half the quantity of* Aqua-Regis, *wherein* Sal Armoniac *hath been dissolved, will do as much as neer double the quantity of that wherein it hath been distill'd,* Aqua-Regis *onely blacks silver, but 'twill slowly dissolve very thin plates of* Copper *and* Tin, *as* Aqua-fortis *will corrode* Leaf-Gold. *But if you draw off the water when it hath dissolved Gold, then 'twill fall upon Silver or other metalls.*

Chap. 42. *T*He ways of making Calcidonies, Jaspers *and* Agats *seems to be the same with making marbled paper described exactly by* Kirch. l. 10. de luce & umbra par. 2. c. 4. *and transcribed by* Schott. par. 1. l. 5. Chrom. 9. *the way whereof is, that several colours are dissolved in several liquors proper to dissolve them, and are such as will not readily or not all mix one with another, when put into water, before they are cast upon the Paper to receive this variety of colours. And so in like manner variety of materials being mixed together, and such as will not incorporate each with other, must needs give various and distinct colours to the metal. Many experiments might be given of tinged liquours, that put into the same Glass, would keep their distinct stations and colours, nay though the liquours were agitated and confounded, they would each return to their proper place and stations. In the preparation of these* Calcedonies, *I shall observe first, that all the*

Observations on the Author. 313

colouring materials, though not all their preparations are used in each of the three ways, viz. Zaffer, Manganese, Silver, Steel, Smalts *and* Lead. *Secondly, the greater variety of ingredients, makes the better* Calcidony, *for the first is more simple than the second, and the second than the third, and our Author commends the last before the second, and that before the first. Thirdly, that some ingredients there are in each of them which contribute no colour at all to the metall, such are* Tartar, Soot, Sal Armoniac, Mercury. *Fourthly, that some of them are of an unctuous nature, as* Lead, Soot, Tartar *and* Smalts, *which may hinder the union of the material one with another, which appears by this, that they do part one from another, and therefore the metall being taken when it begins to grow cold, will then shew some waves, and divers colours very fair,* chap. 42. A great deal of Art there is in working the metall at a due heat, and in the manner also, and in this latter much of Art lyeth as it doth also in marbling Paper.

Porta *teacheth how to colour Glass with various colours, this he found out by chance, when he was making other tryals. Calcined* Tin *takes away the perspicuous colour of Glass and variously colours it, for when 'tis sprinkled by turns on Glasses polished with the wheel, and exposed to a kindled fire, it colours them variously and renders them darker, for one part becomes stone, the other is variously coloured that 'twill seem an* Opal. *But you must often take them out of the fire, and fit them till you have your wish.*

    Here and in many other places our Author tells you that Glass may be wrought into any shape. I shall for the Readers delight set down the most curious I have met with, Card. l. 10. c. 52. de varietate *saw a Cart with two Oxen which was covered with the wing of a Fly,* Agric. l. 12. de re Metall, *saw at* Moran *living Creatures Trees and Ships, and many other famous and admirable works. Master* Howel, pag. 39. *saw a complete Galley, with all her Masts, Sails, Cables, Tackling, Prore, Poop, Fore-castle, Anchors, with her long Boat, all made out in Crystall Glass, as also a man in armor.* Worm. *had in* Musæo, *little statues of Glass, both of men, and other things. The most beautiful Church of Saint* Mark *at* Venice, *adorned within with* Mosaic *work, representing several holy histories with fit colours,*

Observations on the Author. 315

and covered in some places with Gold.

Chap. 48. *P*Ort. l. 6. c. 5. *makes this Amethist colour with a Drachm of Manganese to each pound of Metall.*

Chap. 49. *P*Orta *for the* Saphyre, *adds two Drachms of* Zaphar *to a pound of metall, and the longer (saith he) they continue in the fire, the brighter the colour will be, you must continually mix them.*

Chap. 58. *D*Eep Red *in the original,* rosso in corpo, *whereof thus* Imper. l. 4. c. 1. haver corpo dicono quelli colori che coprono e sono senza transparenza, havere corpo dicono quelli c' hanno transparenza. *The Painters say those colours have body which are close, and without transparency, and those not to have body which have transparency.*

Chap. 61. *G*Lass of Lead, *'tis a thing unpractised by our Furnaces and the reason is, because of the exceeding brittleness thereof. The whole Art of calcining*

316   Observations on the Author.

Lead, to glase their ware withall being the principal thing wherewith that glasing is made, is sufficiently known and practised dayly by the Potters, And could this the be made as tough as that of Crystalline *'twould far surpass it in the glory and beauty of it's colours, of which no man can be ignorant, that hath had any experience of this Metall. That experiment of* Kircher *easily to be tryed & with inconsiderable charge will evince this, thus he,* l. 1. de luce & umbr. par. 3. c. 5. *If you heat with live coals Quick-silver congealed with the vapor of Lead in a Brass-spoon, there will soon appear to you in the melted stuff so great variety of colours, that no greater can be conceived in ihe world, insomuch that none of those which are call'd apparent colours may be compared with them; I remember that trying the reduction of Lead from* Ceruss, *by setting it over the fire, had by putting an Iron sharp pointed into it a considerable quantity of a most brittle matter, not transparent, but adorned with most beautiful colours of* Blew, Green *and* Yellow, *though the later over-ruled both the former and some grains of Lead. I also cast some* Brimstone *into melted* Lead *which became of the fairest skie colour that ever I saw, with the intermixture of other colours*

Observations on the Author.   317

with the Blew, *and those colours not fading but now continuing for these 12 years past,* Libav. de transmut. met. l. 7. c. 20. *saith, that the melters, and tryers of metals daily change* Lead *into Glass, and that this Glass is* Black, Red, Yellow, *or otherwise coloured, as the calcined* Lead *is handled, or as* Lead *is calcin'd into* Lytharge, Ceruss *or* Minium. Quercet, In Hermet. med. defens. cap. 9. *affirms he saw with his own eyes, a Ring made of Glass of* Lead, *which infused in wine a night was a perpetual Purgative, The like variety may be observed from* Bismutum *or* Tin *glass as* <u>Libav</u>. Syntag. Arcan. l.6. c. 4.

    Lead returning into it's body, breaks out the bottom of the pots. *Lead can hardly be so well calcin'd, but some particles thereof will remain uncalcin'd, which the heat of the furnace reduceth to* Lead *again, the same was said of* Ceruss *before, and the like you shall find in* Minium, *the highest calcination used thereof. Now the cause why it breaks out the bottom of the pots seems to be, that receiving there a new calcination, and closing with it's unctuosity, and body the pores of the pots, it hinders the passages of the fire into the metall, which impeded, converts all it's force upon the clay, whereof the pots are made, and by farther calcining it must necessarily make holes in them;* Now that Lead *doth sink into, and not as other metals continue melted on the surface of the pots 'tis manifest by the refiners tests, and Gold-smiths coples, which shew by their colour and weight, what body they have received into them, and by their remelting and reduction of the* Lead. *'Tis true the tests do imbibe some Silver, and therefore they re-melt them in the great heat of the* Almond Furnaces, *and no doubt the same happens to the coples, though the* Essay Masters *of the* Tower *strongly assert the contrary against the importers of* Bullion, But the Essay Masters *at* Gold-smiths-hall *do gain Silver from these coples by melting them down. But in this case some minute parts of Silver onely get into the tests and coples by the mediation of the* Lead *alone, since* Lead *is used in both refining and Essays. But Silver alone nor other metall will at all sink into the test. Another reason of this Accident, may be that the* Lead *insinuating it's self into the pores of the pots, and continued there in fusion, will by getting farther and farther by it's weight into the bottom of the pot at last run out and then leave holes for the metall to follow.*

Our Author mentions not a Jacinth *from Glass of* Lead *nor Glass of* Tin, *nor of* Copper. Bapt. Port. *supplies you with the first,* l. 6. c. 7. *in these words, To make a very* Jacinth, *and not much differing from the true one. Put* Lead *in earthen pots that are very hard in a Glass Furnace, and there let it stand some days, and thus your* Lead *is turned into Glass, and imitates the colour of the* Jacinth. *and of the second* Ib. c. 9. *Melt a pound of* Tin *in an earthen pot that will bear the fire, let it stand in the Glass furnace three or four days, then take out it, and break the vessel, and on the surface you shall find a Glass of a muddy Saffron colour, and if it stand longer in the fire 'twill become more perfect. Neither know we any more perfect in this kind of many we have tryed. But you must put it into the pot well powdered, wherein you must use not onely Mortars and Mills, but the* Porphyrie-stone, *if you would have it lighter, dilute it by adding Glass. Another way reserved for his friends is this, let there be nine parts of calcined* Tin, *seven of* Lead, *two of* Cinnaber, *of* Ferretto *of* Spain, *and of* Tartar *one part and a half, of* Lap. Haematitis *or* Blood-stone, *one part,* Red-ochre *a*

320   Observations on the Author

quarter, do as thou knowest. His Glass from Copper, *l. 6. c. 7. is this, Dissolve Silver, in a strong* Aquafortis, *then* <u>cast the water on</u> Copper-plates, *to which the Silver will stick, which gather and dry, then set it in the Glass furnace, and 'twill be turn'd into an* Emerald *in few days. I commit to you the tryal of other metals, 'tis enough for me to have searched out and shewed the way.*

Chap. 72. B Lew Smalts for Painters. *I cannot find the composition hereof in any writer, but I have been informed by an honest workman in Glass, that 'tis made of* Zaffer, *and* Potashes *calcin'd together in a furnace, made like that for Glass, and that he wrought it in* Germany. *But of this, and all other natural and artificial colour in a treatise designed on this subject.*

Gold hinders the rising of the Metall. *And so doth a little Oyl, or Tallow, thrown into a* Copper *of boyling* Sugar, *hinder it's running over into the fire, though it rise with the greatest fury.*

Chap. 74. T *His way of colouring* Crystall, *teacheth the true and natural way, whereby* Opals, Agats, Jaspers, Chrysolites, Cats-eyes, Marble, &c. *Receive their variety of Colours, they have in themselves, to wit, from exhalations of* Minerals, *supervening to the* præexistent *substance of the stone, as here the colours of* Orpiment, *&c. raised and driven by the heat, penetrate the body of the* Crystall, *and give it this variety of colour. Now if, the matter of the stone being first in liquid form, and therefore capable to receive a tincture, have for it's matrix or womb such a place, whence simple exhalations proceed, the colour is single and unmixt, but if manifold, then the tincture of the stones becomes correspondent to the diversity of the colours arising therefrom. And this appears to be true, by what is frequently observed in larger transparent stones, part whereof will be coloured with their natural colour, and part void of all colour, but simply transparent like Ice. So that the whole stone may well be resembled to frozen water, to that part whereof which was first frozen an accession of colour was made, and none to the other part. Which may be seen more frequently in* Amethists *than in other gems, though many other Jewels afford the like, some having in some part a colour, and some others wholy without any, or else the several parts tinged with diversity of colours;*

322   Observations on the Author.

Chap. 75.   *THe knowledge to imitate Emeralds, &c. There's nothing of value, but some way hath been found to Sophisticate it. And since the counterfeiting of Jewels with exactness, would bring more profit to the inventer, than any other adulteration whatsoever, and perhaps with no real loss to mankind, but great advance, as some Chymists affirm, and therefore not punishable by any law I know of unless in the* Gold smith *who will warrant the counterfeit for true, 'tis no wonder that many means have been to this end and purpose used by pasts, doublets and foils, or colouring the bottom of them, and various other compositions, and artifices, whereof this of Our Author seems the most genuine and natural. Of the fraud in Doublets,* Ferant. Imper. l. 20. c. 14. *gives this relation. A jeweller or Milan sold an Emerald doublet for 9000 Duckats, and the fraud was a long time conceal'd.*

The Chymists *have invented a peculiar though barbarous name for these pasts, and no where extant but amongst themselves. They call them* Amausa, *so* Libav. Joan Isaac, *but* Clauber. Amausæ, *which, whether derived from* Musaicum (*not* Mosaicum *as* Vossius *in his* Glossary, *proves at large) I determine not, though this* Etymon *be very probable, For* Mosaick *work was made in this manner as* Hermol. Barbarus, *describes it.* Musivum opus quod vulgò Museacum vocant, tessulatum lapillis variorum colorum, ex queis arte compositis & coagmentatis omne genus imaginum redditur. *Mosaick work they call that which was checquer'd with stones of divers colours, with which composed and joyn'd by art, all kinds of resemblances are made. These works were anciently made, with small pieces of various* Marbles *of several colours form'd in the shapes of Animals, and sometimes enchac'd with Gold, as appears by* Plin. l. 36. c. 1. Senec. Epist. 86. Philander *in* l. 7. c. 1. Vitruvii *mentions the reliques of some pavements seen by him,* wherein Checquer'd Marbles no bigger than small Beans did accurately and expresly imitate in various colours, the effigies of Fishes and other things. *But the use of coloured Glass succeeded the use of Marbles, and other stones* Libav. *in his* Syntagm. *saith, the* Saracenical *Authors call them* terra Saracenica, *but he confounds these* Enamels *and* Pasts *one with the other. 'Tis true these two are very neer of kin, but are distinguishable by this, that* Pasts

324   Observations on the Author.

are made of Crystall, *prepared and mixed with some Glass, and so wrought into a transparencie, but Enamels have the basis from calcin'd lead and tin, which gives them opacity, corporeity and solidity, by reason of the great quantity thereof mixt with the ingredients.* Glauber *thinks* Furn. Philosop. l. 4. *Pasts were found out by chance by those who reducing calcin'd bodies with a strong fire, converted them into Glass, and adds out of* Isaac Hollandus, *that metalls vitrified and reduc'd yield better and more noble Metalls than those which were first vitrified, to wit, Gold a tincture, silver gold, and copper silver,* &c. *he saith noble Glasses might be made of Metalls, could* Chrysibles *be made strong enough to hold them; what he saith in many words, concerning the preparation of* Crystall *casting into molds and colours, contains nothing but what's vulgar.*

Chap. 76.   Hartm. *in* Praxi Chym. *hath this peculiar way of preparing* Crystall *for making of Jewels. Dissolve, saith he, in water two ounces of purified salt of Tartar, which moisten with* Beechen-ashes *make thereof balls as big as apples; Dry and burn them in a potters furnace in a covered pot, for so the balls will somewhat melt, and stick one to another, let them then be finely pounded & a Lee made thereof, which congele to a Salt. And in this Lee let the* Crystalls *be so often, and so long extinguished, till you can rub them to powder, betwixt your fingers. This being done, let some of the remaining salt be so often purified by solution, coagulation, and calcination, till no fæces at all appear in the solution. Take then of purified salt of* Tartar *two parts, of the foregoing salt prepared,* <u>one part</u>, *melt them together. This will receive all the colours of the whole world, and appears like Oriental gems. Chymical Authors generally prepare* Crystall *this way, onely some extinguish it in Vineger instead of fair water, you may easily know the best way, by the discourse concerning the Glass drops, which is to follow, and doubtless the best way, were to extinguish it often in a strong Lee.*

The making of these Pasts *differs nothing from that of Glass, but that* Pasts *are made of* Crystall *prepared, as the other of* Crystall *metall, the colours in both are the same. And therefore* Porta *calls his Glass tinged with colours by the names of* Amethist, Ruby, &c.

326   Observations on the Author.

Tryal would be made with our English *Diamonds, which are harder and purer than* Crystal.

Chap.77. BOeth. de Boodt, *an excelent writer upon stones, writes thus of adulterating the* Emerald, *This may be done several ways. The best is with* Crystall, Glass *and* Flints *calcin'd, and melted, if a little quantity of* Minium *be added to them. So I have made good ones. He subjoyns, the making of them with burnt* Brass, *half the weight of* Crocus Martis *boil them six hours, and let the pot cool of it's self. If they be well made they will be wholy like those that come from* America. Garcias ab Horto *affirms them to be made fair coloured and very large in* Balaguate *and* Bisnager *of larger fragments of glass pots,* Dalechamp, *thinks some green* Jasper *is to be added to them.* Birelli, l. 8. c. 9. 10, &c. *gives you the same composition with our Author, where you have many more. Another of* Minium *and* Copper-scales, c. 5. *like our Authors,* c. 78. Hartm. *gives several ways, the first obscure, and unintelligible with* Anima Lunæ, *and* Solis, *and* Crystall, *with a little* Sal Armoniac *fixt with lime, a second with* Minium *four ounces,* Crystall *prepared one ounce, gold two drachms; the third with* Brass *calcin'd and powdered fine, mixt with a double weight of Sand for Glass, and standing four days in a very strong fire, and half a day more in a stronger fire. A fourth with his prepared* Crystall *mixt with a little* Copper, *fill herewith a pot half full, set them at a gentle fire four or five hours, then run them in a strong fire, then take away the fire, and break the pot, you shall find the stuff covered with the spume of Lead, which break, and a fair* Emerald *will appear, which he caused to be broke into pieces, and to be cut to his liking. This succeeds not always well, for a serene air is necessary. Therefore he prepared it in a forefold quantity, in four several pots, and so with one labour had four distinct colours one higher than another. For the first he took of* Copper *a scruple, for the second two scruples, for the third one drachm, for the fourth a drachm and a half, and nothing else, for otherwise they will not be transparent. The same is to be done with* Jacinth *and* Topaz, *with* Crocus Martis, *and with the* Saphyre *with* Zaffer.

But in this composition Mars *is wanting to give life and lustre to his* Venus. Card.

328   Observations on the Author.

de var. l. x. c. 52. makes this the colour of stones taken out of the river calcin'd to whiteness and then mixed and melted with an equal quantity of Minium *in a* Chrysible, *and this must be twice done to effect this colour, but this way is wholy insignificant.*

Isaac *affirms of calcin'd* Copperas *and the rest of the* Calces, *that if they be ground with salted water, then washed with fair water, both warm, they will have a far better effect than without these washings, becoming thereby more perfect and fusible.*

Seneca Epist. *91. writes that* Democritus *invented a way to turn stones into* Emerald. *And* Plin. l. 37. c. 12. *saith, that ways are extant in the writings of the Authors, by what means* Emeralds *may be coloured from* Crystalls, *as also other pretious stones, and perhaps differs not from the artifice delivered l.36. c. 26.* de Vitro obsidiano & Myrrhino *of many colours.*

Chap. 81. **B**Apt. Porta *thus adulterates the* Topas. *He mixeth to every pound of Metall a quarter of an ounce of* Crocus Martis, *and a little* Minium, *and*

Observations on the Author.   329

that it may more neatly shine, adds to each pound three ounces of Minium, *but puts in the* Minium *after the* Crocus. Boodt *transcribes this, and adds this also of our Author, and then this following, powder* æs ustum, *native* Cinnaber *and* Crystall, *and four times as much of* Calcined Tin, *set them a day in a fire not too strong, but kept in the same degree, for the said powder easily melts.* Birelli *proceeds this last way, onely changeth* Cinnaber *into* Minium, *and in the very same words, so that* Boodt *had this from him, as the former from* Porta. Hartm. *and* Libav. *with three ounces of ceruss, and Crystall prepared half an ounce. The Author of* quadrig. Chym. *makes salt of* Tin *to be the* Topaz.

Chap. 82.   P*Orta thus imitiates the* Chrysolite, *when you have made a* Topas, *add a little Brass, that it may become more Green, for these two onely differ in this, that the* Chrysolite *shines more neatly,* Claveus *saw silver calcin'd two months in a Glass furnace the twelfth part whereof became a* Citrin *Glass.*

330   Observations on the Author.

Chap. 85. A Saphyre. Glauber *makes this colour with Silver* Marcasite *dissolved in* Aqua-regis, *and precipitated with his liquour from* Flints.

Chap. 90. A Wonderful Red from Gold. *The Chymists with their* menstruums *promise from Gold, a Gold coloured tincture, but I have heard an able* Chymist *offer, not an unconsiderable wager that he would reduce the full quantity of the Gold within few grains (which sure must be lost in the process) when another eminent person of the same profession, had extracted the fullest promised Yellow tincture from it. But the condition was not accepted of. Sure I am that Gold dissolved in* Aqua-regis, *and dropt upon the skin will colour it with a deep purple colour, lasting some days, and this solution poured on a great quantity of water will give it the very same tincture;* Glauber *gives it a fair* Saphyre *colour, being precipitated with a liquour from* Flints. *The tincture of silver is not a skie colour, but white, and for it you have also the undeniable Authority of Master* Boyle *in his Physiological Essays,* pag. 60. *and therefore as I have said before,*

Observations on the Author.   331

the blew must proceed from some Copper *mixt with it.*

Granats of Bohemia. Boeth de Boodt *affirms that these* Granats *from* Bohemia *keep their colour in the fire, but almost all others not, and therefore seem the best for this use, but yet the heat of the Glass furnace consumes it, though it may persist in an ordinary fire.*

Chap. 91.   TAke Ceruss. *Our Author delivers two ways of making* Saccharum Saturni, *the one here of* Ceruss, *the other of* Lytharge, *Chap. 123. onely in this he calcines the* Saccharum, *and out of it calcin'd remakes a new* Saccharum. *The Chymists commonly take* Minium, *some onely calcin'd Lead, all returns to the same purpose, but 'tis observed that* Minium *yields a greater quantity of Salt, and good reason, for that hath had more calcination than any of the other. All make use of distil'd vineger alone, but* Beguin *he substitutes in it's place* Phlegm *of distil'd vineger, but the commentator well passeth a deleatur upon it. Two things I shall here set down, the one that 'tis much better and less chargeable by far, to pour distil'd Vineger on new* Minium *at each time, and not* on that you have used before, for the cheapness of the Minium, and the goodness and quantity of the Saccharum *drawn the first time from the* Minium, *besides the saving a great deal of Vineger this way will advantage the operator much in point of profit. A second thing here to be inserted is a new way, I have not met with in any* Chymical *writer, but invented for my own use, which doth readily and in a moment make it, and I am sure 'tis rather better than worse than the ordinary for* Chirurgical *uses in which I employ it. The manner of making it is this, Take very thin plates of lead, or rather that which hath been long in Glass windows, and dissolve it in Aqua-fortis (good water neer dissolves as much as it's own weight) and the dissolved Lead will soon become a* Saccharum *in the bottom of the Glass. I have in half an hour made a considerable quantity this way in a small glass set in sand and at no great heat, or in a fire shovel over the fire, or in ashes. And certainly this process as more speedy so less expensive, but what this medicine will effect in glass I cannot say.*

Chap. 93.   THis sixth *Book treats of Enamels, which seem to be so named, because 'tis used* in annulis *in rings, or*

Observations on the Author. 333

from the Duch *word* Emailieren *or the French* Esmailler *which comes* à maille macula *a spot as* Minsheu, *for so 'tis laid on. In* Latin Encauston *(that is burnt in, a* καύω *to burn) for so the* Lexicographers *render* Encauston Enamel, Encaustice, *the art of* Enamelling, Encaustes *an Enameler. But the* Encaustum *of the Ancients whereof* Vitruv. l. 7. c. 9. Plin. l. 35. c. xi. Mart. l. 1. &c. *make mention, was a thing quite different from our Enamelling. Concerning which, and the three kinds thereof, see at large* Salmas. *in* Solin. *who truly concludes his discourse, that all this Art is lost.* Porta *makes a* Latin *word, of the Italian* Smalto, *calling them* Smalti *and* Libav. Smalta.

Chap. 94. WHite Enamel, *a new way with* Regulus Antimonii, *you had before,* Libav. & Porta *make it of* Calcin'd Lead *one part, of* calcin'd Tin *two parts, and* Glass *the double.*

Chap. 95. A Turcois, *by* Porta *with* Zaffar *alone.*

334   Observations on the Author.

Chap. 97. **F**Or a Green Porta *takes* aes ustum *which the common people (saith he) call* raminella, *and by our Author* ramina, Chap. 24. *for a deeper colour, and for a lighter, the* Scales *which fall from the hammers, when the* Brass *is hammer'd Red hot.*

Chap. 100. **B**Lack made by Libav. & Porta *with the* Purple *and* Blew *colours, meaning thereby* Manganese *and* Zaffer, *and is the same with our Authors the doses in all of them the same.*

Chap. 103. **R**Ed by Libav. *with* Crocus Martis.

Chap. 108. **A** Lee *of* Barillia *and* Lime. *Much care is to be had of the* Menstruum, *this of* Lime *and* Barillia *are the best, though pot ashes with* Alum, *do very well also. I know an Ingenuous gentleman, who this way hath made all his Colours for plants, which he hath drawn to the life in a large volumne of the most beautiful flours of all sorts in their proper and genuine colour. The vertue of pot ashes (which the*

Observations on the Author. 335

dyers call ware) is seen in their working of Indico *and* Woad *neither of which without these ashes will yield their tincture; for the lightest colours use onely a solution of* Alum *for stronger Salts destroy their colours, as in dying* Soap *ashes, mars the Yellow of Weed or* Fustick, *and in* Chap. 4. Tartar *will not make Yellow in Glass.*

Chap. 110. Whatsoever herb, or flower. *The tryal of our Author is good, but stayning of linnen is a better sign. The rule given by the Merchant to the Mariners in their instructions for forein voyages, is to chaw the plant, and if that colour tinge the spittle deep 'tis good, otherwise not, and so with linnen or fine white paper.*

I shall here give you a catalogue of many plants, &c. *which give a colour, and consequently are fit to make Lakes of, and first those of the dyers, as* Log-wood, *three sorts of* Fusticks *for Yellows, Green, old and young.* Campegiana *and* Sylvester, *which are two sorts of grains or small berries brought from the* West-Indies, *they make a grain colour, though not so good as* Cochineel, *yet they are used instead thereof.* Red-wood, Symach, Brasiletto, *or* Sweet-wood, Turmerick, Safflower, *that is,* Saffron-flower, *but not that of the* Crocus, *but of the* Carthamus *brought from* Italy, Anotto *made of the* Fucus Marinus Tinctorius, *stale and grease, which yields a fair* Scarlet. Weed, *that is,* Genista Tinctoria, *for a Yellow colour.*

Others not used in dying are Saffron, Phalangium Tradescanti, *a very deep and fair* Blew. Cyanus *an excellent Skie for* Dyers. Alga marina Tinctoria *distinct from the former* Fucus, *both mentioned by* Joan. Bauhin. Harebels, *our Purple* Colchicum. *A triplex* Baccifera *a deep Red,* Heliotropium *in whose* juice *rags insuccated make* Turn-sole. Blattaria *with a Blew, and also with a Yellow flower, and the* Convolvulus *narrow leafed of* America; *some plants have a coloured juyce, as the* Spurges, Sow-thistles, Dandelion, Tragopogon, Periplocas, Rampions, Lettices, &c. *most whereof dryed in the Sun turn Yellowish (which makes me suppose* Camboja *may be the juyce of some* Spurge.*) But* Saint Johns *and* Saint Peters Wort, *and* Tutsan *have a reddish juyce in their tops.* Celandine *the greater, and* Felsel Alpini *give a Yellowish juyce. The Berries of many* *plants, also affords colours, as* Dwale garden, Nightshade, *the* Bryonies, Ruscus, Solomons Seal, Herb Christopher, Rasberries, Great-bearing-Cherries, Spina Cervina, *the* Painters Sap-green, Wall-nuts Bezetta, Seu Torna solis Bezedini *of* Wormius *in his* Musæum, l. 2. c. 34. *who thus describes it. 'Tis a fine linnen cloath impregnated with a most Red and Elegant Tincture, But how 'tis prepared, and what is the way of making it, the doner of it* Christopher Herfurt *the Apothecary of King* Christian *the fifth knew not. It seems to be the tincture of* Red-sanders, *wherewith the Cloath is coloured. They use it as* Turnsole *to colour the body and dishes of meat Red: But this is far neater than that, fit for* Cosmeticks, *having this peculiar that steept in water it communicates it's colour thereunto, scarcely to wine, but in no wise to Spirit of Wine, so far he. I have seen this tincture, but made with* Cottonwool, *and 'tis used for a* Fucus, *and common enough with us, and without doubt a singular good Lake might be made therewith.* Amaranthi, balaustia *the seed of* Heliotropium tricoccum *that at first rubbing gives a Green, then a Blew, and lastly a Purple as* Libav. *fragments of the* Alaternus *as* Clus *give a Black,*

Observations on the first Book.

Succory flowers, the flowers of the Scarlet Bean, *of the* Indian Scabious, *the Golden* Marigold *of* Crete, Cerinthe *and* Indian water Cresses, *and many other whereof (God willing) at large hereafter, especially since no Herbarist hath taken notice of the tincture of Plants, nor put them in any tribe, which are of very great use in many* Trades, *and some of those before-mentioned, have been brought into use by* Trades-men; *Leaves that colour, are* stramonium Arbor tinctoria *of* Virginie *whose leaves rub'd on the hands, gives the deepest Green I have seen from any Vegetable, Leaves of* Acanthus *or* Bears-breech. *The true* Tobacco-leaves, *the flowers of* Nigella Hispanica, *which though Blew, being rub'd on the hand, paper or linnen, give a fair Green-colour.*

This way of extracting colours by distillation is now well known and practised, for all the Spirits Chymically drawn, rise white, and they are coloured either by infusion of materials that have tinctures in them, as in the Pharmac. Lond. *the compound Spirit of* Lavender, *and compound* Poppy-water, *and* Aq. Mariae, &c. *and most* Pharmacopæas, *and* Chymists teach this way of our Author.

Observations on the Author. 339

But the extraction of the Spirit of the wine thus tinctured will render the colour dead, and worthless, unless you draw but a little of it, and with no stronger a heat, than that of B.M. for too much heat turns all the colours of vegetables Black, Nay, Lapis Lazuli a hard stone, by too great a heat loseth it's colour.

Chap. 111. T*His preparation I made, and had a dirty Blew therefrom which would do no feat in Pottery. Our Author calls this Blew of* Germany, *and so doth* Birelli, l. xi. c. 106. *onely* Birelli *useth Brimstone, and takes but four parts of Sal Armoniac, you may see many other of this nature there.*

Chap. 112. T*O restore the decai'd colour of* Turcoises. *I doubt and have been told this will not succeed, yet may be much better than those of* Isab. Cortes, l.3. c. 53. *She rubs the stones with* Ultra-marine *that hath stood a day in* Aquafortis, *which being evaporated and dryed, the powder may be used. Secondly, She infuseth them in* Aquafortis, *made of* Vitriol *and* Brass, *then in Vineger, and last of all in water, and each of them some time.*

340   Observations on the Author.

Chap. 113.   A *Mixture to make* Sphears. *Many comcompositions I find in Authors, and because they are of singular use in the Opticks, and nothing published thereof in our own language, I shall here give you such as I have met with. Those* Sphears *or* Glasses *are call'd Metalline, not because they are made of metall, but because some Metalline bodys are mixed with them, and they do as to weight, and appearance much resemble them* Porta. Mag. l. 17. c. 23. *thus prepares the mixture for them. Take a new pot that will bear the fire luted within, dry it twice or thrice, melt therein of* Tartar *and* Crystalline Arsnick *of each two pound, when you see them smoak, put in fifty pound of old worn-brass, melt them six or seven times, that they may be purified and refined, then presently add twenty five pound of* English Tin, *and melt them all together. Take a little hereof with an iron out of the pot, and try whether it be britle or hard, if britle add* Brass, *if hard,* Tin, *or else boil it till some of the* Tin *fly away when it hath the desired temper, cast upon it two ounces of* Borax, *and let it alone till the fume be gone, Then cast it into a mold and let it cool, when cold rub it with a* Pumice *then with* Emerie,

Observations on the Author. 341

when you see the superficies smooth and polished, rub it with Tripoly, *and lastly with, fit* Tin *give it light and lustre, Most add a third part of* Tin *to the* Brass, *that the mass may be harder, and acquire greater perspicuity.*

Porta l. 4. c. 23. *Of his former edition, thus compounds this mixture. 'Tis thus commonly made by all men.* Brass, *and a triple of* Tin, *a little* Arsnick *and* Tartar, *that they may melt, and be incorporated, some add a triple quantity of* Brass *to* Tin *a little* stibium, silver, *and the* White Pyrites; *some make it of* Lead *and a double of* silver, *and 'tis made of other metalls, and otherwise tempered. When they are cast into molds they must be polished and smoothed, that the reflected Ray may bring with it the resemblance of things, and imitate a Looking Glass. Whereunto the smoothness and fitness of the parts much conduceth. If the mixture be not smooth enough, cut or grind it that on one side the image represented may be bigger, and on the other less, and different. If it be rough apply it to the wheel, where arms are polished, and so 'tis burnished. If you make the glass Concave or Convex, lest the motion of the wheel should break the Glass plain a piece of wood, and make it of the shape*

342   Observations on the Author.

of your Glass, and fasten it on with pitch that it stir not. Then rub it over with fine powder of Emery *with a Cloath or Lether, then with fine powder of the* Pumice-stone, *or whilst it sticks to the Wood with Putty (so the* Gold-smiths *call* Tin *calcin'd) mixed with* Tripoly. *And for the last polishing with* Tartar, Soot *and ashes of* Willows *or* Juniper, *which will make it shine best of all.* Emery *is prepared by powdering sercing and wetting.*

Cardan. l. 2. de variet. c. 57. *Glasses call'd* Steel-Glasses *are made of three parts of* Brass, *of one part of* Tin *and* Silver, *and an 18th part of* Antimony. *Most leave out the silver for the charge, others add onely a 24th part, as* Aldrovand. l. 1. c. 4. Musæi Metall *relates. Some make it of a pound of* Tin, *a third of* Brass *melted, and then add an ounce of* Tartar, *and half an ounce of white* Orpiment *all boil'd so long as they smoak. Then they fashion the Molten Metall into the figure of a* Looking-Glass, *on plain tables, heated and dryed with the smoak of* Rosin, *and smoothed with vine ashes, then they afterwards smooth it glewed to Wood with water, and sand, next with* Emery, *or a smooth*

Observations on the Author.   343

Pumice, *thirdly with* Putty, *thus* Cardan, *and from him* Kircher *and* Schwenterus.

Harstoffer. tom. 1. par. 6. q. 13. deliciar. Math. *from* Fliorovant, *takes three quarters of* Tin, *and a quarter of refined* Copper *and melts them, then four ounces of calcin'd* Tartar, Crystalline Antimony *six ounces,* Antimony *sublim'd two ounces, common oyl four ounces,* Marcasite *three ounces; Mix all these, and to every pound of the said metalls, take thereof two ounces, let them evaporate and refine, adding a little* Burgundie-pitch, *when these are consumed pour the stuff in the molds.*

Scal. exerc. 82. Sect. 3. *thus of this mixture, melt nine ounces of* Tin, *three of* Brass, *and then add dryed* Tartar *one ounce, white* Arsnick *half an ounce, let them stand on the fire as long as they smoak, and in the casting, and polishing proceeds as the other Authors.*

Cornæus *communicated to* Schottus *this way. Take ten parts of* Copper, *when 'tis melted, add four parts of* Tin, *then sprinkle a little* Antimony *and* Sal Armoniack, *and stir and mix them till all the dangerous smoak*

344   Observations on the Author.

(from which keep your mouth and nose) vanish then cast it into a mold. I have found (saith he) this mixture by much use to be very good.

Some of these mixtures, and many others like with divers other materials for polishing you may find in Birelli, l. 9. c. 47. *to the* 55. *to whom for brevities sake I refer you.*

Chap. 114.   T*His way of colouring Glass Balls on the inside, is now changed into another of* Pasting Pictures *on the outside of Balls, they are very pleasant, commonly hung up in houses.*

Gesso. *Whereof thus* Cæsalp. l. 1. c. 9. *(the onely Latin Author I find mention it)* est alia terra pallida glebis lapidosis qua utuntur ad Auri-chalcum tergendum, vulgo vocant gessum. *There's another pale earth with stony clots, which they use to scoure* Brass, *they call it* Gessum. *But it seems he knew not what it was, 'Tis a sort of Lime burnt into a pretty hard and very white stony substance, glittering with spots, as Spar doth in* Lead *and* Tin *Ore, and pretty ponderous. To the eye it much resembles* Alabaster, *and is brittle as it, for so is a large piece I have by me.* 'Tis made in Spain, *and carried thence to the* Canary Islands, *and put into the wine transported thence, and gives it a whitish colour and fermentation, and so preserves that wine which would not otherwise keep, but would grow vapid, being transported into other countries.*

Chap. 115. ULtramarine, *so call'd as* Cæsalp. quod forte Egyptum significat aliis prælatum, *this most beautiful colour, and of value equal, if not surpassing Gold, all Authors that treat of stones or colours, deliver the ways of preparing it. 'Tis a very nice colour to make, and unless all the Lapis Lazuli you use be singularly good, all your labour is lost. 'Tis sufficient for me to point at the Authors, who have written of it, omitting their processes because very long and tedious.* Boeth. de Boodt. de gem & Lap. l. 2. c. 123, 124. *to* Chap. 142. *Where he teacheth in a long series of words, to chose the stones (for some of them will bear the fire which* Aldrovand. <u>calls</u> *fixed, others will lose their colour in the fire) then the way to calcine it, to make vessels, Lees, strong and weaker Plaisters, wherewith the colours may be more easily*

346 Observations on the Author.

drawn forth, and how it must be washed to serve for Pictures. And in the last Chapter *he teacheth a shorter and less expensive way to extract this colour. Next him followeth* Birelli, *who somewhat shorter delivers all these Processes,* l. 9. *from* Chap. 80. *to* Chap. 109. *Some painters onely grind the* Lapis Lazuli *into a fine powder, and so use it.*

Chap. 116. Lake from Cochineel. *No doubt this word comes from the Gum call'd* Lacca, *the colour and tincture whereof have both the same colour, with this of the* Painters. Math. in l. 1. Diosc. c. 23. *asserts there are many kinds of Artificiai Lake which are made of the Sediment of several tinctures. One is made of the Berry (head) of Burnet which they commonly call* Cremese *and* Cremesino (Crimson) *another of* Chermes Berries, *a third of true* Gum-lacc, *and lastly a fourth of* Brasil, *the worst of all, but he sheweth not the way of making either of them.*

Concerning this place, and the mistakes of Math. *herein, hereafter in a Treatise designed for colours,* Birell. l. 11. c. 39. *teacheth a way to make a Lake of this* Gum. *Take (saith he)*

Observations on the Author. 347

about twenty pound of mens urine, which boil and scum well, put a pound of Gum-lacc, *and five ounces of* Alum *into it, set them over the fire. Boil them till the colour be extracted, make proof with a little of it, then add of* Alumen Saccharinum, *what quantity you judge fit, then strein it as the other Lakes are.*

I find in several writers receits for making Lakes, differring onely, either in the materials from which, or in respect of the Menstruum *wherewith they are extracted. Some use* Chermes-berries *or* Grains *(a sirup whereof the Apothecaries have of a noble tincture) and they are gathered from the* Ilex *thence call'd* Coccigera, *a tree whereof you may see in a garden in* Old-street, London, *neer the* Pest-house, *but it never bore fruit in England, another grew in his Majesties* Privy garden *at* White-hall, *but 'twas lately cut down, by the ignorant usurpers. Some use the* Cochineel, *which is a Maggot or fly bred on the* Ficus Indica, *whereof see at large,* Joan de Laet descript. Ind. l. 5 c. 3. *as also* Herrera & Zimenes. *Others use dyed* Flox *(the most common) which our Author here teacheth how to die, and this is the best way. Others take the Scowrings of Cloath dyed in Stammel or Scarlet.*

348   Observations on the Author

Hernandez *in his* Hist. l. 3. chap. 45. *thus of making* Lake *in the* Indies. *Of* Nocheztli, *that is* Cochineel, *sometimes a* Purple, *sometimes a* Scarlet *colour is made, according to the various ways of preparing it.*

The most exquisite is made by heating it with the water of the dicoction of the tree call'd Totzuatl, *adding* Alum, *and the setling is form'd into* Cakes.

As for the Menstruums *they are* Lees *made by our Author of* Vine *or* Willow *or of other soft Wood. Others make it of* Oaken *or other strong ashes, yet the Lee must be no stronger than being put upon the tongue, 'twill prick or bite it a little onely. Surely* Aqua-fortis *might do very well, since we see it so far advanceth the colour of* Cochineel *in our incomparable* Bow-dyes. *The only inconvenience in* Lakes *hence made would be, that they would soon Tarnish and lose their colour in the air, or with wet, by reason of the Salts relenting; but perhaps this might be remedied by extracting and washing of these Salts without any damage to the colour. Now all writers proceed the same way in discharging the colour, precipitating streyning and drying the Lake made. As to the last I shall add this, that* Chalk-stones *sooner dry by imbibing the moisture than Bricks do, as the constant practice of Painters in making Pastils, and of the Refiners in drying their* Verditers *confirmeth. Before the Lake be fully dry, they form it into Balls or cut it with a wooden Knife (not with an Iron one) into what shapes and figures they please, or they may do as Painters for their Pastils, cast them in furrows made in the stone.*

Chap.117. S Aline *of the* Levant, *with my Author* Pilatro di Levante; *this word* Pilatro *I cannot find in any Italian writer, this exposition of the word I had from an ancient person who wrought at* Moran, *he added 'twas a Salt extracted from the froath of the Sea, coagulated through the extreme heat of the countrey. The name of* Saline, *and this way of generation thereof I have had from other workmen, but the exposition from him alone.*

350   Observations on the Author.

Chap. 118. *B*Irelli makes his Lake from Brasil *thus, He first extracts a tincture from* Flox, *and then takes a pound of* Brasil *cut (ground is better) and boils the* Lee *to the consumption of a fingers thickness, then streins it, and adds to the streyned liquor one ounce of Gum* Arabick *in powder, and reboils it, and boils away half as much as before, then mixeth both the liquours with a stick, then proceeds with the* Hippocras.bag, &c. *as before.*

Chap. 124. A Fair rose Red Rosichiero, *which* Porta, l. 6. c.9. *calls* Rosa-clerum, *& teacheth this way of making it. Put 10 pound of* Crystall *into a pot, when 'tis well melted, put in a pound of the best* Minium *by halfs at a time, stir them speedily, then with Iron ladles cast them into water, and that thrice, then mix five ounces of calcin'd* Brass *and* Cinnaber *of the deepest Colour, and having stirr'd them well, let them settle three hours. When you have so done superadd of Glass of* Tin *three ounces, mix them without intermission and you shall see in the Glass the most* Florid *colour of the* Rose, *which you may use to* Enamel *upon Gold.*

Chap. 126. *T*O *fix* Sulphur *he teacheth another* Chap.129. *way, Another Process to the same purpose, but much larger,* Birel *delivers,* l. 1. c. 50. *But* Sulphur *thus prepared will easily rise sublim'd* with Sal Armoniack. *None that I have met with affirms such a fixation of* Sulphur, *as* Helmont *doth, for in his mixture of Elements, he saith, he knew ways whereby whatsoever* Sulphur *was once dissolved might be fixed into a* Terrestrial *powder. Our Author no where mentions any use of this powder in the Art of Glass.*

Chap. 129. *A Transparent Red.* Libav. l. 2. Tract. 1. c. 35. *By Conjecture hits right on this colour from Gold in these words, I judge that from a red tincture of Gold dissolved into a liquour or oyl, and especially with* Crystal, *a* Rubie *may not unfitly be made. Of which conjecture he assigns this reason, because* Rubies *are frequent where Gold is found, and therefore 'tis consentaneous that gold there doth degenerate into this jewel.*

Chap. 131. *To make vitriolum Veneris, Glaub. l.2. Furn. Philosop. proposeth this short way.* Spirit *of* Sal Armoniac *powred on calcin'd Copper, made by frequent ignition and extinction, in an hours space extracts a Blew colour, which when dissolved, decant off, and set in a cold place, and 'twill yield a most elegant Blew* Vitriol. Croll. *in his* Basil. Chym. *describes well the making of this Medicine.* Beguin, c. 17. *sets down this way Powder calcin'd Copper, or it's scales very fine, which digest 24 hours in distil'd vineger. Pour out the Tinctured Vineger by inclination, and pour on more till 'twill be no more coloured.* Filtre *the decanted liquours, Evaporate, or distil off a third part, set the remainder in a cold place, and you shall have Green and obscure* Vitriolum Veneris.

FINIS.

AN ACCOUNT OF THE GLASSDROPS.

THese Drops were first brought into *England* by His Highness Prince *Rupert* out of *Germany*, and shewed to his Majesty, who communicated them to His Society at *Gresham* College. A Committee was appointed forthwith by the Society, who gave this following Account of them as 'tis Registred in their Book appointed for that purpose, and thence transcribed by their permission, and here published. The which I the rather desired, that this might be a pattern for

354 *An account of the Glass Drops.*

experiments to be made in any kinde whatsoever, as being done with exceeding exactness.

This account was given to the Society by *Sir* Robert Moray MDCLXI.

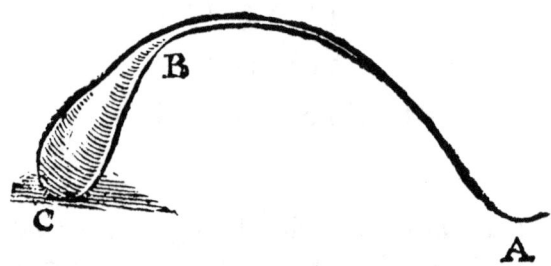

A B the thread, B C the body, B the neck,
A the point or end of the thread.

They are made of *Green-glass* well refined; till the Metall (as they call it) be well refined, they do not at all succeed, but crack and break, soon after they are drop't into the water.

An account of the Glass Drops.

The best way of making them, is to take up some of the Metall out of the pot upon the end of an Iron rod, and immediately let it drop into cold water, and there lye till it cool.

If the Metall be too hot when it drops into the water, the Glass drop certainly frosts and cracks all over, and falls to pieces in the water.

Every one that Cracks not in the water; and lies in it, till it be quite cold, is sure to be good.

The most expert Workmen, know not the just temper of heat, that is requisite, and therefore cannot promise before hand to make one that shall prove good, and many of them miscarry in the making sometimes two or three or more for one that hits.

Some of them frost but the body falls not into pieces; others break into pieces before the red heat be quite over, and with a small noise; other soon after the red heat is over, and with a great noise; some neither break nor crack, till they seem to be quite cold; others keep whole whilest they are in the water, and fly to pieces of themselves with a smart noise

356 *An account of the Glass Drops.*

as soon as they are taken out of the water; some an hour after, others keep whole some days or weeks, and then break without being touched.

If one of them be snatched out of the water whilst it is red hot, the small part of the neck, and so much of the thred or string it hangs by, as has been in the water, will upon breaking fall into small parts, but not the Body, although it have as large cavities in it, as those that fly in pieces.

If one of them be cooled in the air, hanging at a thread, or on the ground, it becomes like other Glass, in all respects, as solidity, &c.

When a Glass drop falls into the water, it makes a little hissing noise, the body of it continues red a pretty while, and there proceed from it many eruptions like sparkles, that crack, and make it leap up and move, and many bubbles do arise from it in the water, every where about it, till it cool: but if the water be ten or twelve Inches deep, these bubles diminish so in the ascending, that they vanish before they attain the superficies of the water; where nothing is to be observed, but a little thin steam.

An account of the Glass Drops.

The outside of the Glass drop is close and smooth like other Glass, but within it is spungious, and full of Cavities or Blebs.

The figure of it is roundish at the bottom for the most part, not unlike a pear pearl, it terminates in a long neck, so that never any of them are straight, and most of them are Crooked and bowed into small folds and wreaths from the beginning of the neck till it end in a small point.

Almost all those that are made in water have a little proturberance or knob a little above the largest part of the body, and most commonly placed on the side towards which the neck ends, although sometimes it be upon that side that lies uppermost in the vessel where it is made.

If a Glass drop be let fall into water scalding hot, it will be sure to crack and break in the water either before the red heat be over, or soon after.

In Sallet Oyl they do not miscarry so frequently as in cold water.

In oyl they produce a greater number of bubbles, and larger ones, and they

An account of the Glass Drops.

bubble in oyl longer than in water.

Those that are made in oyl have not so many, nor so large blebs in them, as those made in water, and divers of them are smooth all over, and want those little knobs that the others have.

Some part of the neck of those that are made in oyl, & that part of the small thread that is quenched in it cool'd, breaks like common Glass. But if the neck be broken neer the body, and the body held close in ones hand, it will crack and break all over: but flies not into so small parts, nor with so smart a force and noise as those made in water, and the pieces will hold together till they be parted; and then there appears long streaks or rays upon them pointing towards the center or middle of the body, and thwarting the little blebs or cavities of it, whereof the number is not so great, nor the size so large as in those made in water; if the Glass drops be dropt into vineger, they frost and crack, so as they are sure to fall to pieces before they be cold, the noise of falling in is more hissing than in water, but the bubbles not so remarkable.

In milk they make no noise, nor any

An account of the Glass Drops.

bubbles that can be perceived, and never miss to frost and crack, and fall in pieces before they be cold.

In spirit of wine they bubble more than in any of the other liquors, and while they remain entire, tumble too and fro, and are more agitated than in other liquors, and never fail to crack and fall in pieces.

By that time five or six are dropt into the spirit of wine, it will be set on flame: but receive, no particular taste from them.

In water wherein *Nitre* or *Sal Armoniack* hath been dissolved, they succeed no better than in vineger.

In oyl of *Turpentine* one of them broke, as in the spirit of wine, but the Second set it on fire, so as it could no more be used.

In *Quick-silver*, being forced to sink with a stick, it grew flat and rough on the upper side: but the experiment could not be perfected, because it could not be kept under till it cool'd.

In an experiment made in a Cylindrical Glass, like a beaker filled with cold water, of seven or eight onely one succeeded, the

360 *An account of the Glass Drops.*

rest all cracking and breaking into pieces, onely some of the company, who taking the Glass in their hand, as soon as the drop was let fall into it, observed that at the first falling in, and for some time after, whilst the red heat lasted, red sparks were shot forth from the drops into the water, and that at the instant of the eruption of those particles, and of the bubbles which manifestly break out of it into the water, it not only cracks and sometimes with considerable noise, but the body moves and leaps, as well of those that remain whole in the water, as those that break.

A blow with a small hammer, or other hard tool will not break one of the Glass Drops made in water, if it be touched no where but on the body.

Break <u>off</u> the tip of it, and it will fly immediately into very minute parts with a smart force, and noise, and these parts will easily crumble into a coarse dust.

If it be broken, so that the sparks of it may have liberty to fly every way, they will disperse themselves in an orb, with violence like a little Granado.

An account of the Glass Drops.

Some being rubed upon a dry tyle, fly into pieces by that time the bottom is a little flatted, others not till half be rub'd off. One being rub'd till about half was ground away, and then layed aside did a little while after, fly in pieces without being touched. Another rub'd almost to the very neck on a stone with water and *Emery* did not fly at all.

If one of them be broken in ones hand under water, it strikes the hand more smartly, and with a more brisk noise than in the air: yea, though it be held near the superficies, none of the small parts will fly out of it, but all fall down without dispersing as they do in the Air. One of them broken in Master *Boyles* Engine, when the Receiver is well Evacuated will fly in pieces as in the open air.

Anneal one of them in the fire, and it will become like ordinary Glass, onely the spring of it is so weakned, that it will not bend so much without breaking, as before.

A Glass drop being fastned into a cement all but a part of time neck, and then the tip of it broken off, it made a pretty smart noise, but not so great as those use to do that are broken in the hand, and though it clearly appears to be all shiver'd within, and the colours turned grayish, the outside remained smooth, though cracked, and being taken in pieces, the parts of it rise in flakes, some Conical in shape, and so crack all over, that it easily crumbled to dust.

One fastned in a ball of cement some half an Inch in thickness, upon the breaking off the tip of it, it broke the ball in pieces like a Granado.

Two or three of them sent to a *Lapidary* to peirce them thorow, as they do Pearls, no sooner had the tool entred into them, but they flew in pieces as they use to do when the tip of them is broken off.

FINIS.

An Appendix.

In the Chapter of the Furnaces I gave an account of the Instruments used about Crystalline Metall, but having omitted there those which are used in making Green Glasses take them here as they follow.

TWo Bars *to lift their pots into the Furnaces, each neer four yards long.*

A Padle, *to stir and move the Ashes and sand in the Calcar.*

Rakes *to rake the Ashes and Sand to and fro in the Calcar.*

Procers *are Irons hooked at the extremity to settle the Pots in their places, whether set too far or near, or on either side from the working hole.*

Ladles *to empty out the Metall from one Pot into another whether the Pots break, or to any other purpose.*

Small Ladles *for each Master workman to scum the* Sandever, *and dross, from the pot wherein he worketh.*

Strocals *a long Iron instrument like a Fire-shovel to carry the Metall out of a broken into a whole Pot.*

An Appendix.

Forks *to prick betwixt the bars of the Fire place to help the descent of the ashes that the fire may burn clear, and bright.*

Sleepers *are the great iron bars crossing smaller ones which hinder the passing of the coals, but give passage to the descent of the ashes.*

Ferrets *are the Irons wherewith they try whether the Metall be fit to work, as also those irons which make the Ring at the mouth of Glass Bottles.*

Fascets *are Irons thrust into the bottle to carry them to anneal.*

The Pipes *are the hollow irons to blow the Glass.*

Ponte *is the Iron to stick the Glass at the bottom for the more convenient fashioning the neck of it.*

Pontee stake *is the Iron whereon the Servitors place the irons from the Masters when they have knock't off the broken pieces of Glass.*

Cassia stake *is that Iron whereon lyeth a piece of wood, on which wood they lay the Glass when they have taken it off the pipes, & whereon they turn the Glass to fasten the Pontee to it.*

Shears *are the Instruments to form and fashion the Glass.*

Scissers *cut the Glass, and even it.*

An Appendix.

Cranny *is a round Iron whereon they roul the Glass to make the neck of it small.*

Tower *is the Iron on which they rest their Pontee when they scald the Glass.*

Several sorts of Iron Molds *wherein they make their works of several figures, protuberances, &c. according as they are cut in them.*

FINIS.

Author's Errata & Corrigenda.
(Merrett's pagination)

EPist. Ded. *read* pour on you, *p. 12. line. 2. r.* from although to the end at the latter end of Chapt. 3. *p.16. l. 13. &c. r.* refine the Glass, *ib. l. 16. r.* is made. *p.24. l.14. r.* 10. *p.106. l. 15. r.* lead again. *p.159 l. 11. for* Cochin *r.* Blew. *p. 207. l.8. r.* Borint. *p.210 1. 16. r.* I sod. *p.211. l.13. r.* Belluac. *p. 271. l. 17. r.* that make. *p. 314. l.4. r.* cast the water on.

Additional corrections in this edition
(These are underlined in the text)

pp. 39, 40 & 57, read beads *instead of* counting houses.

p. 74, l. 4, read incorporated *not* incoporated; *l. 6 read* 1601 *not* 1691.

p. 156, Ch. CVI, read Ch. 23 *instead of* 21 *and* Ch. 22 *instead of* 23.

p. 173, Ch. CXVII, l. 4, insert just.

p. 175, l. 3 read too *not* two.

p. 176, l. 3 from bottom, read haply *for* happily.

p. 177, last l. read your *not* you.

p. 199, l. 4. read off *not* oft

p.207, l. 5 from bottom, read falsly *for* fasly.

p.215, l. 2. read restitutionis *for* rectitutionis.

p. 216, l. 13 read wetted *for* witted; *4 from bottom read* Nomenclatore *for* no Menclatore.

p. 224, l. 3 from bottom, read both *not* hoth.

p. 240, l. 14, read ash-hole *for* ashole.

p. 245, l. 1, insert are made.

p. 247, l. 17, read off *not* of.

p. 254, l. 4, insert h *in samphire.*

p. 267, para 2, l. 1, insert that.

p. 291, l. 5, read discourse *for* discours.

p. 299, l. 6 from bottom, read use *not* used.

p. 300, l, 9 from bottom, read exsiceateth *not* exiccateth

p. 304, l. 4, read nor *for* or; *l. 3 from bottom, read* off *not* of.

p. 307, l. 11, read Antimony *for* Antionomy.

p. 315, chap. 58, read Red *for* Reed *and* transparenza *for* trasparenza.

p. 317, l. 13, read Libav. *for* Libar.

p. 325, ll. 10-11, read one part *not* part one.

p. 345, chap. 115, read calls *not* cals.

p. 360, l. 6 from bottom, read off *not* of.

p.363, l. 13, read to *not* too.

GENERAL INDEX
Page numbers are those at the bottom of the pages
Eccentricity of spelling is ignored

A

Abraham, 47, 275
Acanthus, 396
Agate, 117, 119, 131, 142, 271, 370, 378
Agricola, 12, 23, 29, 31, 267, 301, 325, 337
Alabaster, 402
Alchemist, 47
Alchemy, 8, 35
Alexandria, 268
Alicante, 13
Alpinus, 309, 310
Alum, 120, 124, 219, 230, 234, 236–7, 241, 268, 293, 405
Alumina, 14
Amethyst, 145, 150, 157, 373, 379, 383
America, 279, 281, 381, 394
Amsterdam, 18, 31, 33, 35
Anima saturni, 242
Annealing, 3, 24, 26–7, 30, 131, 247, 272, 293, 294, 299, 301–12, 304, 306, 359–60, 362, 419, 422
Antimony, 134, 140, 175, 177, 338, 365, 391, 400–1
Antwerp, 7, 8, 108, 142, 178, 186, 199, 229
Apuleius, 265–6
Aqua fortis, 24, 46, 91, 92, 117, 120, 123–26, 128, 133, 135, 136, 139, 140, 141, 142, 178, 186, 199, 229, 273, 338, 351, 367

Aqua Regis, 24, 46, 92, 117, 126, 259, 273
Aqua vitae, 221, 222, 237
Arabian, 100, 274, 313, 333, 408
Aristotle, 8, 278
Arno, river, 65, 76, 121
Arsenic, 121, 140, 224, 367, 398, 399, 401
Art of Glass, 3, 4, 5, 18, 20, 22–4, 33, 37, 39, 41, 43, 49, 59, 86, 99, 120, 218, 258, 349, 409
Ashes, 9, 11, 12, 23, 48, 59, 60, 62, 68, 69, 71, 73, 112, 135, 141, 160, 223, 233, 244, 251, 256, 268–9, 298–9, 308–12, 314, 318, 321–31, 333–4, 347, 362, 390, 392–3, 400, 406, 421–2
Azure, 75, 112, 130, 134, 208, 217, 257

B

Babel, 276
Babylon, 15
Balass, 163, 176–8, 241, 260
Barillia, 13, 48, 59, 60, 75–7, 98, 218, 259, 282, 312, 330, 392
Beads, 22, 97, 98, 115, 287, 424
Beans, 73, 381, 396
Beech, 325, 382
Beer, 25, 267, 274, 284
Beguin, 369, 389, 410
Belus, river, 47, 318–9
Birelli, 23, 384, 387, 397, 402, 404

427

Biringuccio, V., 23
Black, 15, 58, 61, 65, 80–81, 83, 86, 88, 94, 96, 100, 105, 114, 127, 137, 145, 152, 153, 154, 158, 159, 188, 198, 210, 211, 212, 227, 229, 233, 252, 253, 255, 258, 260, 297, 337, 340–1, 345–8, 354–5, 364, 369, 375, 395, 397
Bladder, 333
Blancourt, H. de, 34, 36
Blue, 60, 86, 101, 103, 107, 109–11, 114, 115, 129–30, 134, 175, 191, 217, 222, 223, 225, 230, 254–6, 258, 260, 265–6, 287, 314, 336–8, 340, 342, 345–7, 349, 350, 351, 354–57, 361, 365, 374–5, 389, 392, 394–7, 410.
Bollito, 59, 65
Borage, 221
Borax, 398
Botany, 23
Box, 321
Boyle, Hon. Robert, 6, 20, 21, 287, 388, 419
Brabant, 142
Brazier, 104
Bramble bush, 73, 323
Brass, 48, 58, 93, 95, 100, 101, 104, 107, 108, 112, 133, 140, 158, 170, 172, 174, 184, 194, 201, 207, 208, 210, 213–5, 224, 232, 252–3, 259, 289–90, 314, 338, 339, 347–8, 350–4, 357, 360–2, 364–6, 374, 384–5, 387, 392, 397–402, 408
Brimstone, 87, 88, 89, 92, 108, 128, 130, 140, 246, 252, 253, 263, 374, 397,
Broom, 217–9
Burnet, 111, 404

C

Cabbage, 221
Caesalpinus, 23, 336
Caesar, 265, 289–90, 292
Calcar, 9, 12, 13, 37, 64, 75, 77, 105, 114, 243, 282, 297–8, 302, 324, 329, 331, 357, 365, 421
Calcination, 7, 13, 15, 16, 17, 25, 28, 58, 61, 66, 75, 77, 82, 83, 87, 88, 89, 93, 94, 95, 100, 104, 118, 138, 140, 153, 155–6, 158, 160, 163–6, 168, 171–2, 174, 183–5, 197–8, 200–1, 204–5, 207–8, 210, 213–5, 227, 238, 240, 243–5, 247–8, 250, 252–5, 259, 273, 300, 303, 314–6, 318, 331, 333, 336, 338, 348, 353–54, 356–9, 363, 365–6, 371, 373, 375–8, 382–7, 389, 391, 400–1, 403, 408, 410
Calx, 153, 204–5, 248
Canary, 304, 403
Caput mortuum, 107, 113, 140, 170
Carbuncle, 275
Cardanus, 23, 320, 336–7, 342, 400–1
Carnation, 221
Carrera, 65
Casino, 8, 132
Cats-eyes, 215, 378
Cavalet, 90, 301
Ceruss, 21, 134, 196, 325, 374–5, 387, 389
Chalcedony, 8, 92, 95, 117, 118, 120, 128, 129, 131, 132, 135, 136, 137, 141, 143, 183, 260, 355, 370–1
Charles II, 20, 323
Chemical art, 52
Chrysolite, 178, 181, 190, 198, 260, 378, 387

Cinnabar, 134, 139, 199–200
Clay, 26, 29, 30–1, 302–304, 331, 359, 375
Coal, 3, 15, 27, 29, 88, 90, 94, 95, 112, 122, 123, 127, 138, 164, 176, 178, 183, 252, 297–8, 300, 357, 359, 361–2, 369, 374, 422
Conciator, 60, 67, 79, 97, 102, 115, 298, 332, 334
Cobalt, 11
Cochineal, 7, 230–3, 236–8, 260, 393, 404–6
Coleworts, 73
Collet, 305, 331, 335
Conterie, 2, 22
Copper, 7, 11, 58, 87, 93, 104, 124, 133, 134, 224, 240, 244–5, 247–8, 252–55, 314, 316, 338–9, 345, 347, 350–64, 368, 370, 377–8, 392, 384–6, 389, 401, 410
Copperas, 88, 314, 316, 345, 350–2, 354–5, 363, 368, 386
Cornwall, 360
Cortes, Isabella, 367, 397
Cotton wool, 395
Counting houses, 2, 22, 97, 227
Crete, 396
Crocus Ferri, 89
Crocus Martis, 17, 89, 90, 91, 92, 93, 109, 110, 111, 112, 113, 128, 130, 133, 135, 139, 141, 169, 174, 186–8, 190, 209–11, 244, 248–9, 259, 346–7, 349, 362, 364, 366, 384–6, 392
Crocus metallorum, 281
Crucibles, 90, 94, 138, 160, 175–8, 181, 183–4, 190, 198, 227, 252, 260, 382, 386

Crystal glass, 10, 37, 49, 64, 65, 66, 67, 68, 70, 72, 73, 74, 78, 79, 80, 81, 96, 98, 101, 110, 111, 112, 116, 129, 135, 146, 149, 150, 152–3, 155–8, 160–1, 166, 171–8, 183–93, 200, 205, 239, 241, 243–4, 248, 250, 257–60, 274–8, 282, 293, 301, 304, 311, 315–6, 318, 321, 330, 356, 372, 374, 378–9, 382–7, 398, 401, 408–9, 421
Cullet, 29

D

Decolorizing, 9
Descartes, 287
Duncan, G.S., 7, 33

E

Egypt, 13, 279–80, 282–3. 309–12, 348, 403
Egg, 122
Egg, philosophical, 46
Egmarine, 181, 365–6
Elmfield, 3
Emerald, 17, 100, 112, 113, 125
Emery, 398
Enamel, 10, 11, 15, 26, 31–2, 203–17, 242–43, 248–9, 251, 261–2, 349, 365, 367, 381–2, 390–1, 408
Encyclopaedia Brittanica, 37
Esau, 47

F

Fairlight, 28
Fallopius, 267–8
Fern, 71, 320
Fernando, Duke, 136
Ferretto, 87, 88, 133, 139, 209, 211,

259, 349, 377
Fir, 325
Fire, 8, 16, 18, 19, 24, 25, 27, 29, 30, 45, 48, 51, 55, 58, 61, 63, 65, 66, 68, 69, 77, 81, 83, 88, 94, 96, 99, 104, 105, 119, 122, 123, 128, 131, 135, 136, 138, 142, 148, 164–5, 175–6, 178, 183, 185–6, 196–7, 200, 203–5, 209–10, 213, 218–9, 223, 225, 228, 231–3, 238, 243, 246, 252, 253–4, 256–7, 264, 266–9, 271–2, 274, 277, 279–83, 295, 298–9, 300, 302–4, 307, 317, 321, 331, 333, 334–5, 338, 342, 343–344, 347, 352–4, 356–7, 359–60, 362, 366, 368–9, 372–5, 377–8, 382, 385, 387, 389, 398, 401, 403, 405, 417, 419, 421–2
Flanders, 7, 136
Florence, 6, 7, 44, 65, 98, 132
Florentine [coin], 87, 108
Fornello, 100, 114
Frankfurt, 34
Frankincense, 227, 275
Frederick II, 36
Frisius, A., 31, 33–5
Frit, 9, 12, 14, 49, 64, 66, 67, 70, 73, 74, 77, 78, 79, 82, 96, 98, 101, 102, 103, 105, 129, 135, 146, 148, 150–60, 166, 168, 171–4, 205, 241, 243–4, 247–8, 259, 297–8, 302, 329–31, 348, 365
Furnaces, 11, 12, 29, 47, 68, 69, 95, 118, 122, 129, 130, 135, 139, 148–9, 297–307, 317, 328, 330, 332–4, 336, 338, 347–8, 356–63, 365–6, 373, 375–8, 382, 387, 389, 421

G

Garlic, 357
Garnet [Granat], 145, 149, 163, 172–3, 194, 260, 388
Gasparetto, A., 27
Geissler, F., 34, 35
Genista, 323, 394
German(y), 8, 20, 24, 34–5, 337, 341, 378, 397, 411
Gesso, 226, 402
Ghirodolpho, P., 8, 142
Gillyflower, 221
Glauber, 382, 388
Gold, 46, 48, 50, 51, 70, 72, 126, 145–7, 214, 217–8, 243–4, 248, 250–1, 261, 274–6, 287–90, 292, 295, 353, 370, 273, 376, 378, 381, 382, 385, 388, 403, 409
Goldsmith, 15, 87, 203, 205, 251, 376, 380, 400
Grain-colour, 102, 103
Green glass, 28, 30, 110, 112, 274, 282, 301–2, 304–5, 318, 334–5, 421
Gresham College, 6, 20, 41, 54, 264, 411

H

Haematite, 377
Harvey, W., 19, 22
Hawthorn, 321, 323, 325
Helmont, 270, 282, 315, 409
Heliotrope, 394–5
Hermes trimegistus, 283
Hierom [Jerome], 47, 274
Holbach, Baron d', 34
Holland, 329, 360
Hollandus, I., 15, 181–2, 195, 382

Holly, 321

I

Iron, 11, 24, 58, 81, 85, 86, 90, 92, 96, 112, 124, 134, 140, 153, 158–9, 165–7, 184, 194, 208, 252, 253, 272, 281, 298–302, 304–6, 311–2, 316, 335, 338, 341–3, 346–7, 351, 355, 358–9, 363, 364, 369, 374, 398, 407, 408, 413, 421–3
Italy, 56

J

Jacinth, 377, 385
January, 142
Jasper, 117, 119, 131, 271, 378, 384
Job, 47
Juniper, 333, 400
June, 71

K

Kali, 48, 282, 294, 309–12, 321, 323, 325, 328–9
Kelp, 282, 321–2, 328
Kidney, 333
Kiln, 16, 165–6, 204, 280–1
Kircher, 23, 279, 283, 285, 374, 401
Kunckel, J., 24, 30, 33–6

L

Ladle, 13, 62, 85, 86, 96, 99, 253, 256, 298, 334, 408, 421
Ladroni, 342
Landiamo, N., 8, 132
Lapis calamanaris, 93, 338–9, 351–2, 357
Lapis lazuli, 163, 174–5, 218, 227–30, 260, 338, 397, 403, 404

Latin, 2, 6, 12, 18, 31–3, 35, 391, 402
Lattimo, 15, 155, 157, 174, 261
Lavender, 396
Lead, 15–8, 21, 25, 28, 155, 164–5, 168, 171, 173–4, 188, 194, 197, 204, 242, 244–5, 248, 348–9, 374–7, 390–1, 399
Lead glass, 15, 17, 28, 31, 37, 163–4, 168, 247, 259, 260, 261, 375
Lead oxide, 11,
Leek, 353
Leer, 30, 90
Lees, 49, 61, 62, 68, 69, 78, 309, 314–6, 327, 331, 342, 403, 406
Leipzig, 34
Levant, 59, 68, 72, 74, 75, 98
Libavius, 8, 23, 290, 301, 307, 375, 380–1, 387, 391–2, 395, 409
Liguria, 86
Lime, 14, 74, 121, 122
Linen, 122–3, 233, 235, 284, 342, 393, 395–6
Linseed, 227, 228
Lion-colour, 129
Litharge, 242, 389
Lode-stone, 337, 340–2
London, 5, 17–8, 19–20, 27–8, 33, 36, 319, 405
Looking-glass, 275, 284, 399–400
Lucretius, 279, 296

M

Madder, 217, 236
Magnesia, 14, 340
Maidstone, 28, 319
Malleable glass, 24, 48, 283, 293–6
Mallow, 221

431

Manganese, 11, 29, 55, 58, 79, 80, 86, 96, 106, 108, 110, 112, 128, 134, 140, 146, 148–52, 154–6, 160, 172–3, 191–4, 201, 206–7, 210–3, 215, 238–9, 241, 244–5, 259, 320, 326, 331–2, 336–8, 340–7, 371, 373, 392
Marble, 16, 29, 65, 145, 157, 165, 167
Marigold, 396
Maurach, H., 36
May, 71
Medici, Lord Don A., 7, 43
Medicine, 45, 130
Mediterranean, 328
Mercury, 24, 133, 136, 138, 271, 274, 286, 293, 371
Merrett, C., 4, 5, 6, 7, 14, 18, 20, 22, 23, 28, 29, 30, 31, 32, 33, 35, 37, 97, 115
Microscope, 295
Milan, 33
Millet, 73
Minium, 189–93, 374, 384–7, 389–90, 408
Minories, 27
Moon, 71, 286
Moray, Sir Robert, 20, 412
Mosaic, 15, 372, 381
Murano, 56, 65, 86, 98, 308, 347, 372, 407

N

Neile, P., 26
Neri, A., 3, 4, 5, 7, 9, 10, 12, 13, 14, 18, 24, 29, 31, 33, 37, 56
Nero, 292
Newcastle, 30, 317
New England, 322
Nile, 312

Nimenes, E., 8, 142
Nimrud, 15
Nitre, 125, 269, 315, 325–8, 330, 363, 367, 368, 417
Nonsuch, 30, 31, 304, 359
Normandy, 28
Nurnberg, 34

O

Oak, 123, 251, 325, 327, 332, 406
October, 4, 6, 321
Onion, 357
Opal, 163, 177–8, 260, 372, 378
Opticks, 285
Occhio, 90, 100, 104, 243, 301
Orange, Prince of, 143
Orpello, 93
Orpiment, 140, 175, 177, 179, 226, 271, 379, 400
Oxford, 18, 324

P

Padua, 22
Parkin, M., 2
Paris, 6, 34–6
Paste, 26, 41, 50, 55, 121, 136, 181–2, 184–8, 191, 194–5, 199–201, 227–30, 239, 243, 260–1, 349, 364, 380–3
Peacock, 90
Pear, 321, 415
Pearl, 145, 161, 260, 415, 420
Pelican, 46
Penshaw, 30
Phillips, Sir T., 33
Philander, 381
Piedmont, 86, 146, 239
Pimpernel, 221

Pine, 227, 322
Pisa, 7, 65, 71, 99, 104, 107, 110, 151, 207, 235–9
Pliny, 23, 47, 265, 289–90, 292–4, 326, 330, 333, 341
Plutarch, 279, 333
Poggendorf, J.C., 7, 8
Poison, 14, 270
Polverine, 13, 14, 17, 37, 59, 60, 61, 62, 68, 70, 72, 74, 77, 78, 81, 82, 108, 111, 112–3, 153, 160, 166, 168, 171, 200, 243, 259, 282, 308, 310, 312, 315, 326, 330, 336
Pompeii, 27
Poppy, 220–1, 396
Porphyry, 85, 94, 184–5, 200, 222, 227, 230, 238, 239, 377
Portuguese, 8, 142
Porta, B. della, 23, 285, 301, 366, 383, 386–7, 391–2, 398, 408
Poland, 322
Potash, 14
Pots, 15–6, 26, 30–1, 60, 79, 81, 87, 96, 102, 105, 108, 114, 138, 147, 149–50, 153, 155–9, 164–9, 171–2, 174, 176, 185–6, 200, 205–6, 208, 210, 212–4, 237, 240–1, 243–5, 247–8, 255, 271, 280–1, 298–300, 304–7, 309, 320–2, 329, 331–5, 337, 344, 347–8, 352, 359–60, 362, 366, 369, 374–8, 382, 384–5, 392, 397–8, 408, 413, 421
Prince Rupert, 20, 21, 411
Pumice, 281, 398, 400–1
Purple, 90, 217, 241, 252, 260, 337, 340–2, 346, 388, 392, 394–5, 406
Putty, 400–1
Pyrites, 345, 349, 399

Q
Quicksilver,120, 128, 137, 222, 374, 417

R
Radcliffe, 27
Ravenscroft, G., 3, 17, 18, 28, 37
Red, 11, 61, 83, 89, 90, 92, 95, 99, 103, 123, 130, 133, 134, 145, 146–7, 159, 176, 194, 197, 212, 217–8, 220, 253, 255, 261, 271–2, 298, 305, 311, 314–5, 346, 362–3, 368, 373, 388, 392, 395, 408–9, 413–5, 418
Red-lead, 113, 129, 134, 140, 186–8, 197–9, 225, 245
Red-hot, 24, 85
Reeds, 73
Reigate, 28
Retort, 26, 46, 238–9, 257–8, 368
Reverberation, 9, 16, 30, 88, 95, 111, 166, 238, 282, 297, 299, 358, 363
Ricci, 288
Rochetta, 13, 48, 59, 60, 75, 76, 81, 98, 106, 107, 116, 147, 149, 151, 154, 156, 166, 259, 308–9
Rock crystal, 9, 31
Rome, 285, 289
Rosichiero, 117, 119, 408
Royal Society, 5, 6, 18, 19, 20, 21, 26, 27, 35
—— College of Physicians, 4, 18
Ruby, 123, 163, 176–8, 195, 239, 250, 409
Rushes, 73
Russia, 322

S

Sal ammoniac, 92, 119, 126, 128, 133, 134, 139–40, 175, 177, 222, 238–9, 271, 369

Salt, 9, 23, 49, 61, 63, 69, 70, 74, 76, 77, 109, 114, 115, 120–1, 124, 131, 137–8, 160, 196–8, 200, 205, 219–20, 224, 238, 243, 259, 267–70, 273, 294–6, 298, 308–16, 320–31, 334–6, 345, 353, 355–7, 363, 365, 368–9, 382–3, 386–7, 389, 393, 406–7

Salicornia, 309–10

Salt, common, 121, 137, 138, 335

Saltpeter, 120, 124, 221, 238, 294–5, 315, 325–6, 335, 355, 365, 369

Sand, 9, 60, 68–9, 76–7, 129, 147, 196–8, 223, 238, 254, 256, 267–70, 272, 295–6, 298, 303, 316, 318–20, 325–6, 328–31, 334–5, 337–9, 343, 345–6, 385, 390, 395, 400, 421

Sandever, 80, 270, 298, 326, 329, 334–5, 343, 421

Sapphire, 145, 151–2, 163, 173, 181, 191, 192, 195, 198, 201, 261, 340, 365, 373, 385, 388

Scummer, 13, 61, 63, 316

Sea-green, 93, 96, 98, 101, 104, 105, 106, 107, 109, 114, 116, 130, 171, 201, 214, 258, 350

Seneca, 386

Serce, 16, 36, 66, 72–3, 76, 84, 88, 90–2, 94–5, 100, 105, 148, 166, 168, 171–4, 198, 200, 204–6, 229, 232, 252–3, 255, 318, 331, 336, 358

Seraveza, 65

Silver, 46, 48, 120, 126, 128, 136, 138, 139, 194, 222, 288–90, 292, 338, 352–3, 369–71, 376, 378, 382, 387–8, 399–400

Sky-colour, 93, 99, 114, 115, 163, 227, 356, 374, 388, 394

Smalt, 129, 134, 175, 191, 225, 230, 371, 378, 391

Smoke, 3, 11, 12, 51, 127, 176, 253, 297–8, 302, 311, 398, 400–1

Soap, 309–10, 314, 345, 349, 393

Somerset, 347, 357

Spain, 13, 59, 75, 87–8, 98, 106, 128, 133, 139, 209, 211, 259, 275, 282, 312, 348–9, 351, 377, 403

Spagyrical art, 45, 52, 108

SSS, 87, 88, 95

Stourbridge, 31

Steel, 15, 48, 65, 89–92, 153, 158, 159, 224, 318, 347, 351, 364, 371

Strabo, 315

Suarez y Nunez, M.G., 34

Sulphur, 46, 119, 125, 133, 198–200, 245–6, 251, 253, 255, 259, 353–4, 356, 364, 409

Summer, 25, 73, 274, 312, 321–2, 326, 356

Sun, 71, 91, 200, 274, 284, 286, 312, 322, 356, 394

Syria, 13, 47, 59, 283, 309

T

Tamarisk, 279, 333

Tarso, 9–12, 14, 37, 60, 65, 67, 70, 72–3, 76, 77, 82, 84, 147, 150, 183, 205, 243, 317

Tartar, 11, 13, 61–2, 64, 72, 82–4, 117, 119, 127, 130, 135, 141, 146, 148, 150. 154, 161, 201, 205, 211, 213, 224,

231–2, 240, 243, 245, 247–9, 300, 314–5, 345, 371, 377, 382–3, 393, 398–9, 400–1
Taste, 10, 25, 48, 60, 68, 196, 269, 273, 310–1, 324, 326, 329, 334, 417
Terebinthine, 344
Terrestriety, 62, 69, 70, 76, 84–85, 102, 118, 126, 138, 175, 197–9, 234
Tesino, 65
Thorpe, W.A., 17
Tiberius, 24, 48, 283, 289
Tin, 15, 48, 139, 153, 155–6, 158, 204, 224, 243, 248, 316, 371, 375, 377, 387
Tin oxide, 11
Titania, 14
Tobacco, 30, 303, 312, 323, 396
Tools, 23, 27, 31, 32, 34, 35, 346, 418, 420
Topaz, 163, 171, 176–8, 182, 189, 198, 201, 261, 385, 386, 387
Tradescanti, 394
Trebian, 219
Tremolante, 93
Tripoli, 309, 399–400
Turner, W.E.S., 3,
Turpentine, 227, 304, 345, 417
Turquoise, 100, 114–5, 217, 223, 260–1, 391, 397
Tuscany, 65, 71, 76, 86

U

Ultramarine, 8, 139, 217–8, 227, 229–30, 260, 397
Urine, 363, 405

V

Vauxhall, 27
Velvet, 145, 152–3, 210, 212
Venice, 10, 14, 33, 86, 227, 270, 342, 349, 367, 372
Verdigrease, 134, 140, 186, 194, 226, 314, 350–1, 356, 364
Verucola, 65
Vine, 233, 310, 406
Vinegar, 17, 85, 135, 137, 139, 141, 169, 186–7, 190, 196–7, 210–1, 259, 312, 351, 353, 363–4, 383, 389–90, 397, 410, 416–7,
Violet, 151, 151, 173, 191–3, 215, 220–1, 261
Virginia, 323, 396
Vitriol, 88, 124, 125, 140
Vitriol of Venus, 107, 108, 113, 196, 198
Vitruvius, 265, 381, 391
Volterra, 114

W

West Indies, 393
Wheat, 13, 64
Willow, 233, 400, 406
Wine, 11, 22, 25, 61, 83, 127, 146, 219, 224, 243, 245, 274, 283–4, 307, 312, 314, 315, 336, 343, 351, 375, 395, 397, 403, 417
Winter, 278, 321, 326
Woad 266
Wood, 11, 29, 58, 76, 78, 103, 122, 123, 137, 186–7, 226, 233, 279, 306, 321, 323, 325, 332–3, 357, 393–4, 399–400, 406–7, 422
Wood, F.J. 1
Wool, 121, 230, 237, 265

Woolwich, 21, 27–8, 319, 331
Worcestershire, 19, 31

Y
Yellow, 95, 102, 125, 126, 130, 145–8, 158, 163, 165–6, 171, 174–6, 197, 199, 201, 213–4, 217–20, 241, 261, 278, 344–5, 350–1, 374–5, 388, 393, 394

Z
Zaffer, 11, 24, 29, 85–6, 97, 98, 101, 106–7, 128, 134, 139, 149–52, 172–3, 190–4, 201, 207–8, 210–1, 214, 258–9, 331–2, 336–40, 346–7, 371, 378, 385, 391, 392
Zanech, 47

www.ingramcontent.com/pod-product-compliance
Lightning Source LLC
Chambersburg PA
CBHW071434300426
44114CB00013B/1435